100 Years of General Relativity – Vol. 2

CHERN–SIMONS (SUPER)GRAVITY

100 Years of General Relativity
ISSN: 2424-8223

Series Editor: Abhay Ashtekar *(Pennsylvania State University, USA)*

This series is to publish about two dozen excellent monographs written by top-notch authors from the international gravitational community covering various aspects of the field, ranging from mathematical general relativity through observational ramifications in cosmology, relativistic astrophysics and gravitational waves, to quantum aspects.

---

100 Years of General Relativity – Vol. 2

CHERN–SIMONS (SUPER)GRAVITY

Mokhtar Hassaine
Universidad de Talca, Chile

Jorge Zanelli
Centro de Estudios Científicos, Chile

W⊖ World Scientific

NEW JERSEY · LONDON · SINGAPORE · BEIJING · SHANGHAI · HONG KONG · TAIPEI · CHENNAI · TOKYO

Published by

World Scientific Publishing Co. Pte. Ltd.

5 Toh Tuck Link, Singapore 596224

USA office: 27 Warren Street, Suite 401-402, Hackensack, NJ 07601

UK office: 57 Shelton Street, Covent Garden, London WC2H 9HE

British Library Cataloguing-in-Publication Data

A catalogue record for this book is available from the British Library.

100 Years of General Relativity — Vol. 2
CHERN–SIMONS (SUPER)GRAVITY

ISBN 978-981-4730-93-8

In-house Editor: Song Yu

Typeset by Stallion Press
Email: enquiries@stallionpress.com

Printed in Singapore

Preface

This book grew out of a set of lecture notes on gravitational Chern-Simons (**CS**) theories, developed over the past decade for several schools and different audiences of graduate students and researchers. Those notes were circulated on the High Energy arXiv [1] receiving several comments, corrections and additions, until the last version of hep-th/0502193 in 2008 [2].

Our interest in CS theories originated in the possibility of constructing gauge-invariant theories that could include gravity consistently. CS gravities are only defined in odd dimension and are a very special class of theories in the Lovelock family[1] that admit local supersymmetric extensions. In those theories, supersymmetry is an off-shell symmetry of the action as in a standard gauge theory [3–5].

From a theoretical point of view, CS forms correspond to a generalized gauge-invariant coupling between connections — like the Maxwell and Yang-Mills fields — and charged sources: point-like (particles) or extended (membranes). The simplest example of such interactions is the familiar minimal coupling between the electromagnetic potential (abelian $U(1)$ connection) and a point charge (0-brane). More generally, a $(2n+1)$-CS form provides a natural gauge-invariant coupling between a charged $2n$-brane and a nonabelian connection one-form. This is true for any gauge connection and in particular for a connection of the relevant algebras in gravity and supergravity.

Most CS theories are useful classical or semiclassical systems and although many aspects of quantum CS systems have been elucidated, the full quantization of CS field theories in dimensions greater than three is still poorly understood. There is no gauge-invariant perturbative scheme and no natural notion of energy so the usual assumptions of quantum field theory, like the existence of a stable vacuum,

---

[1] The Lovelock gravitation theories are the natural extensions of General Relativity for dimensions greater than four (see Chapter 4). They yield second order field equations for the metric, describing the same degrees of freedom as Einstein's theory in a way that is invariant under general coordinate transformations and under a local Lorentz transformations.

are not guaranteed. Nevertheless, apart from the arguments of mathematical elegance and beauty, the gravitational CS actions are exceptionally endowed with physical attributes that suggest the viability of a quantum interpretation. CS theories are gauge-invariant, scale-invariant and background independent; they have no dimensionful coupling constants, all constants in the Lagrangian are fixed rational coefficients that cannot be adjusted without destroying gauge invariance. This exceptional status of CS systems makes them classically interesting to study, and quantum mechanically intriguing yet promising.

Acknowledgment It is a pleasure for us to acknowledge a long list of collaborators, colleagues and students who have taught us a lot and helped us in understanding many subtleties about geometry, Chern-Simons theories, supersymmetry and supergravity in all these years. We have benefited from many enlightening discussions with all of them and some have worked through the manuscript and corrected many misprints and suggested changes to improve the notes. Our thanks go to P. D. Alvarez, L. Alvarez-Gaumé, A. Anabalón, R. Aros, A. Ashtekar, R. Baeza, M. Bañados, G. Barnich, M. Bravo-Gaete, C. Bunster, F. Canfora, L. Castellani, O. Chandía, J. A. de Azcárraga, N. Deruelle, S. Deser, J. Edelstein, E. Frodden, A. Garbarz, G. Giribet, A. Gomberoff, J. Gomis, M. Henneaux, L. Huerta, R. Jackiw, J. M. F. Labastida, J. Maldacena, C. Martínez, O. Mišković, P. Mora, S. Mukhi, C. Núñez, R. Olea, P. Pais, S. Paycha, V. Rivelles, E. Rodríguez, P. Salgado-Arias, P. Salgado-Rebolledo, A. Sen, G. Silva, S. Theisen, F. Toppan, P. Townsend, R. Troncoso, M. Valenzuela and B. Zwiebach.

This work would not exist without the understanding and patient support of our families and friends. Our warm thanks for the continued support and collaboration go to our colleagues and staff at CECs-Valdivia and at University of Talca.

This work was supported in part by FONDECYT grants 1140155 and 1130423. CECS is funded by the Chilean Government through the Centers of Excellence Base Funding Program of Conicyt.

Contents

Chapter 1

The Quantum Gravity Puzzle

As is well-known, three of the four fundamental interactions of nature — electromagnetism, weak and strong interactions — are accurately described as Yang-Mills (**YM**) theories in the Standard Model, our best current description of the microscopic world [6]. YM theories are based on the principle of gauge invariance, according to which the interaction results from the existence of a locally realized symmetry, which requires a correlation between particles at different locations in spacetime. This correlation appears as an interaction that we interpret as a force field, and it is this particular form of the YM interactions that makes them consistent with quantum mechanics.

Gravitation, the fourth fundamental interaction, described by Einstein's General Relativity (**GR**), is our best current understanding of the universe at large. GR is also based on a gauge principle, but cannot be quantized following naively the same steps that succeed in the YM case. What is most puzzling — and frustrating — about this problem is that both YM interactions and gravitation possess very similar gauge symmetries, yet the former make remarkably successful quantum theories while the latter does not.

The basic building blocks of nuclei and atoms — leptons and quarks — are spin-1/2 particles described by fields that belong to the *fundamental* (vector) representation of the gauge group G. The interactions that bind leptons and quarks together are mediated by spin-1 particles, the photon, the intermediate vector bosons and the gluons. These carriers of interactions are represented by connection fields that transform in the *adjoint* representation of the same group G. The SM is the mathematical framework describing how all this comes about.

GR describes the gravitational interaction as the effect of the spacetime curvature on the motion of free bodies (free fall) and also describes how matter and energy deform the spacetime geometry. The bases of Einstein's construction are two fundamental postulates, the Principle of Covariance and the Principle of Equivalence. The first is the statement that it is possible to write down the laws of physics in such a way that they make sense in any reference frame, inertial or not. It means that the laws of nature should preserve their meaning under changes of reference frame. It must be possible to express the laws in a way that retains their form under arbitrary changes of spacetime coordinates, they must be *generally covariant*.

The second postulate can be stated as follows: in the neighborhood of every spacetime point it is always possible to pass to a reference frame which is locally indistinguishable from an inertial frame. In practice this means that no matter how curved spacetime might be, it can always be locally approximated by a flat spacetime (Minkowski space), where physical laws are invariant under Lorentz transformations.

These two postulates together mean that all laws of physics should be invariant under Lorentz transformations acting locally. Hence, GR is implicitly a gauge theory after all, where the gauge group is $SO(3, 1)$ (Lorentz group) and Einstein should be credited as the first to conceive a nonabelian gauge theory, preceding the notion of gauge invariance itself, long before the seminal work of Yang and Mills [7]. The gravitational analogue of the electromagnetic vector potential A_μ is the Lorentz connection $\omega^a{}_{b\mu}$. While the quantum of the vector potential is the photon, one could expect the quantum of the Lorentz connection to be the graviton, but without a quantum theory of gravity this issue remains obscure. Even if we don't know how to express gravity as a quantum theory, the fact that it could be formulated as a gauge theory seems like an important feature.

1.1 Renormalizability and the Triumph of Gauge Theory

1.1.1 *Quantum field theory*

The Standard Model of high energy physics is a remarkably successful, predictive and precise theory for particle interactions that describes three of the four fundamental forces of nature with enormous accuracy. The dynamical structure of the model is built on the assumption that nature is invariant under a group of local transformations, which is acting independently at each point of spacetime. This requirement drastically restricts the type of couplings among the fields describing the basic constituents of matter and the carriers of their interactions.

A most important feature of the standard model is the way in which it avoids inconsistencies: renormalizability, absence of anomalies, etc. In fact, it is this feature that makes it a reliable tool and at the same time is the prime criterion for model building. The remarkable thing is that renormalizability and lack of anomalies are so restrictive that only a very limited number of possible actions pass the test. This is reassuring since it means that, at a fundamental level, nature is not arbitrary, but it is dictated by a symmetry principle, which sounds almost like an aesthetic principle of mystical origin. The opposite extreme, on the other hand, would be rather embarrassing: having a large a number of physically sensible and equally acceptable theories, why didn't nature choose any of the other theories?

Although not all gauge-invariant theories are guaranteed to be renormalizable, those that describe the Standard Model are renormalizable gauge theories. This is an unexpected bonus of the gauge principle, since gauge invariance was not introduced to cure the renormalizability problem, but rather as a systematic way to bring about interactions that would respect a given symmetry. Thus, gauge invariance seems to be a crucial ingredient in the construction of physically testable (renormalizable) theories. Symmetry principles, then, are not only useful in constructing the right (classical) action functionals, but they are often sufficient to ensure the viability of a quantum theory built from a given classical action. An important feature that explains in part the usefulness of gauge invariance in quantum theories is the fact that the symmetry does not depend on the field configurations. Gauge invariance is a property of the action, not a feature of a solution and not even a property of a class of solutions. This means that gauge invariance is a consequence of the type of fields involved and the way they occur in the action, independently of whether they satisfy the classical field equations or not. Gauge symmetry is an off-shell property of the system, and not just a feature of its classical manifestations.

Now, if gauge invariance implies relations among the divergences appearing in the scattering amplitudes in such a way that they can be absorbed by a redefinition of the parameters in the action at a certain order in the semiclassical (loop) expansion, it should do the same at all orders since the symmetry is not spoiled by quantum corrections. In contrast, symmetries that are only realized if the fields obey some classical equations of motion something often referred to as an on-shell symmetry are not necessarily respected by a quantum theory.

The underlying structure of the gauge principle is mathematically captured through the concept of fiber bundle. This is a systematic way of implementing a group acting locally on a set of fields carrying a particular representation of the group. For a discussion of the physical applications, see Refs. [8, 9].

1.1.2 *Enter gravity*

Attempts to set up a perturbative expansion to describe the gravitational interaction as one would do with a standard quantum field theory have consistently failed. Progress in this direction has remained rather formal, and after more than 80 years of efforts, no completely satisfactory quantum theory of gravity in four dimensions, analogous to QED or QCD, is known.[1]

The straightforward perturbative approach to quantum gravity was explored by many authors, starting with Feynman [13] and it was soon realized that the

---

[1]In three spacetime dimensions, the analogue of Einstein gravity admits a quantum description even though this theory has no local propagating degrees of freedom [10]. In four dimensions, the representation of gravity in terms of loops –loop quantum gravity– offers a promising alternative [11], although it is unclear how to make contact with classical GR. Here we do not discuss this important approach and refer the reader to one of the many available texts on the subject [12].

perturbative expansion is nonrenormalizable (see, e.g., Ref. [14]). One way to see this is that if one splits the gravitational Lagrangian as a kinetic term plus interaction, the coupling constant for graviton interactions is Newton's constant G, which has dimensions of $mass^{-2}$ in four dimensions. This means that the diagrams with greater number of vertices require higher powers of momentum in the numerators to compensate the dimensions, and this in turn means that one can expect ultraviolet divergences of all powers to be present in the perturbative expansion. The lesson one learns from this frustrating exercise is that General Relativity in its most naive interpretation as an ordinary field theory for the metric is, at best, an effective theory. For an interesting discussion of how one can live with an effective theory of gravity, see Ref. [15].

The situation with quantum gravity is particularly irritating because we have been led to think that the gravitational interaction is a fundamental one. The dynamical equations of complex systems, like viscous fluids and dispersive media, are not amenable to a variational description and therefore one should not expect to have a quantum theory for them because they are not truly fundamental. The gravitational field on the other hand, is governed by the Einstein equations, which are derived from an action principle. Hence, it is natural to expect to find a well-defined path integral for the gravitational field, even if calculating amplitudes with it could be highly nontrivial [16].

General Relativity seems to be the only consistent framework for gravitational phenomena, compatible with the principle that physics should be insensitive to the observer's state of motion. This principle is formally translated as invariance under general coordinate transformations, or general covariance. This invariance is a local symmetry, analogous to the gauge invariance of the other three fundamental forces of nature and one could naively expect that this gauge symmetry would play the same role as in the quantization of other three interactions. Unfortunately, general relativity does not qualify as a gauge theory, except for the remarkable accident mentioned in three-dimensional spacetime, and for a generalization of this case to all odd dimensions, as discussed in Chapters 4 and 5.

The difference between a coordinate transformation and a proper gauge transformation is manifest in the way they act on the fields. Gauge transformations transform the fields without changing the spacetime point,

$$\text{Gauge:}\quad \psi^a(x) \longrightarrow \psi^{'a}(x) = M^a_b(x)\psi^b(x), \tag{1.1}$$

where $M^a_b(x)$ is an element of the gauge group acting locally. Coordinate diffeomorphisms, on the other hand, change the field components as well as the arguments,

$$\text{Diffeomorphism:}\quad \psi^a(x) \longrightarrow \psi^{'a}(x') = M^a_b(x)\psi^b(x), \tag{1.2}$$

where x'^μ are functions of x. It is possible to define the action of an infinitesimal coordinate transformation on some appropriate combination of fields that do not change the argument. This can be achieved using a Taylor expansion to rewrite

(1.2) as

$$\psi^a(x) \longrightarrow \psi'^a(x) = M_b^a(x)\psi^b(x) - \xi^\mu \partial_\mu \psi^a(x)$$
$$= \tilde{M}_b^a(x)\psi^b(x) - \partial_\mu[\xi^\mu \psi^a(x)], \qquad (1.3)$$

where $\xi^\mu \equiv x'^\mu - x^\mu$, and $\tilde{M}_b^a(x) = M_b^a(x) + (\partial_\mu \xi^\mu)\delta_b^a$, with $M_b^a(x) = \delta_b^a + \partial_b \xi^a$.

In order for this to be a useful local symmetry, the Lagrangian of a physical system should be such that it changes by a total derivative under (1.3). Under such transformations, the four-dimensional gravitational action changes by a term that vanishes only if the field equations hold, meaning that this is at best an *on-shell symmetry*. In the three-dimensional case, on the other hand, the Einstein-Hilbert action (possibly supplemented by a cosmological constant term) is invariant under the transformations of the form (1.3), without requiring additional on-shell conditions. One way to understand this is to observe that the diffeomorphisms in three-dimensional gravity are contained in the $ISO(2,1)$ (or $SO(2,2)$, or $SO(3,1)$) gauge symmetry of the theory [17]. This "miracle", however, does not occur in higher dimensions.

1.2 Minimal Coupling, Connections and Gauge Symmetry

Consider a physical system invariant under a set of continuous rigid transformations forming a Lie group G. By rigid we mean that the parameters that define each transformation do not depend on the spacetime coordinates. A typical example could be a rigid rotation by a given angle θ around a certain axis of all the spins in a magnetic system. The parameter θ can take any value in the range $0 \le \theta \le 2\pi$, and the axis of rotation $\hat{n}$ can point in any direction in space and a rigid action means that the axis or rotation and the magnitude of θ is the same for all spins in the system. In other words, if φ represents the dynamical variables and $I[\varphi]$ is the action, then

$$I[\varphi^{(\hat{n},\theta)}] = I[\varphi],$$

where $\varphi^{(\hat{n},\theta)}$ represents φ rotated by an angle θ around $\hat{n}$.

As shown in a general context by Utiyama [18], a continuous rigid symmetry of a physical system can always be made into a local one in which the transformation parameters depend on the position in spacetime. In the previous example this would mean taking the angle and the axis of rotation as functions of the position, $\theta(x)$ and $\hat{n}(x)$. In order to turn a rigid symmetry into a local one, some adjustments must be made in the action of the system. A spacetime vector field $A_\mu(x)$ can be brought in, to define the following derivative operator,

$$\nabla_\mu = \partial_\mu - iA_\mu. \qquad (1.4)$$

The requirement that the derivative $\nabla_\mu \varphi$ transform in the same vector representation as φ is achieved by demanding that $A_\mu(x)$ be in the adjoint representation
of the Lie group G. Then, ∇_μ is said to be a covariant derivative for the group
G (see below for the expressions of the transformations (1.14), (2.10)). As is well
known, replacing the partial derivative ∂_μ by ∇_μ in the original action $I[\varphi]$ makes
the system invariant under the same group G acting locally (G-gauge invariant).
In this way a coupling between the original field φ and the vector field $A_\mu(x)$ is
introduced and it can be shown that the equations that govern φ as well as the
interaction are also gauge invariant.

The kinetic term for the vector field $A_\mu(x)$, however, is not necessarily fixed by
the local symmetry. A common — and extremely useful — choice of kinetic term
for the vector field $A_\mu(x)$ consists of demanding that the Lagrangian be a gauge
invariant function of the field strength $F_{\mu\nu} = [\nabla_\mu, \nabla_\nu]$, namely

$$L_0(A) = -\frac{1}{4} Tr \left[F^{\mu\nu} F_{\mu\nu} \right]. \qquad (1.5)$$

This is the choice that governs electrodynamics and chromodynamics, abelian and
nonabelian Yang-Mills theories, respectively. In this case, the gauge symmetry also
significantly restricts the dynamics of the vector potential itself by fixing the form
of its propagator.

The quantum operator corresponding to the spacetime vector field A_μ describes
the particles responsible for the interactions like the photon in the electromagnetic case, the massive meson fields for the weak interaction or the colored
gluon fields of the strong interaction. The number of degrees of freedom and
the multiplets of fermions to which A_μ couples depends on the particular Lie
group $G(x)$ that is being gauged. It is remarkable how this simple symmetry
requirement naturally introduces these fields with all of their phenomenological
implications. In other words, the promotion of a global symmetry to a local one
naturally introduces an extra vector field whose dynamics also enjoys the same local
symmetry.

The occurrence of this local symmetry also has an important mathematical coun-
terpart in the theory of fiber bundles, where the fields that transform in irreducible
representations of the Lie group (the matter fields) define sections of the principal
fiber bundle, and the vector fields A_μ are the so-called connections. The covariant
derivative (1.4), determined by this connection, can be viewed as the quantum version of the minimal substitution of classical electrodynamics, $p_\mu \to p_\mu - eA_\mu$, as if
the formalism dictated by gauge invariance were tailor-made for the transition to
quantum mechanics. It is remarkable that this little trick is ideally-suited to the
needs of relativistic, non-relativistic, classical and quantum theories. It is a clear
sign of the robustness of the gauge principle in physics.

Historically, the idea of gauge symmetry can be traced to Hermann Weyl's seminal paper of 1918 [19] whose motivations were two-fold: First, to construct a local
scale symmetry so that the magnitudes of vectors would be path-dependent on

equal footing with their orientations. Second, he was hoping to provide a unifying theory including gravity and the only other interaction known at the time, the electromagnetism. Let us briefly recall the main aspects of Weyl's idea.

In a gravitational context, spacetime is a pseudo-Riemannian manifold endowed with a metric $g_{\mu\nu}$. The covariant derivative is defined by the Levi-Civita connection as

$$(\nabla_\mu)^\alpha_\nu = \delta^\alpha_\nu \partial_\mu + \Gamma^\alpha_{\mu\nu}, \tag{1.6}$$

where

$$\Gamma^\alpha_{\mu\nu} = \frac{1}{2} g^{\alpha\sigma} \left(\partial_\mu g_{\sigma\nu} + \partial_\nu g_{\sigma\mu} - \partial_\sigma g_{\mu\nu} \right), \tag{1.7}$$

are the Christoffel symbols, also denoted sometimes as $\{^\alpha_{\mu\nu}\}$. Under an continuous and differentiable coordinate change $x^\alpha \to y^\alpha = y^\alpha(x)$, the Christoffel symbols transform as

$$\Gamma^\lambda_{\mu\nu}(x) \to \Gamma'^\lambda_{\mu\nu}(y) = \frac{\partial y^\alpha}{\partial x^\mu} \frac{\partial y^\beta}{\partial x^\nu} \frac{\partial x^\lambda}{\partial y^\gamma} \Gamma^\gamma_{\alpha\beta} + \frac{\partial^2 y^\alpha}{\partial x^\mu \partial x^\nu} \frac{\partial x^\lambda}{\partial y^\alpha}. \tag{1.8}$$

In complete analogy with the electromagnetic case, where the field strength is defined by the commutator of covariant derivatives (1.4), the covariant derivative operator (1.6) now defines the Riemann curvature tensor,

$$R^\alpha_{\mu\nu\beta} \equiv [\nabla_\mu, \nabla_\nu]^\alpha_\beta \tag{1.9}$$

$$= \partial_\mu \Gamma^\alpha_{\nu\beta} + \Gamma^\alpha_{\mu\lambda} \Gamma^\lambda_{\nu\beta} - (\mu \leftrightarrow \nu), \tag{1.10}$$

which measures the infinitesimal rotation of a vector under parallel transport generated by ∇ on an infinitesimal loop defined by the directions dx^μ and dx^ν.

Levi-Civita [20] noticed that the covariance of the operator ∇_ν and of the curvature tensor are consequences of the transformation law of the Christoffel symbols (1.8), and not necessarily because Γ is constructed from the metric tensor. In fact, any three-index array $\hat{\Gamma}^\alpha_{\mu\nu}$ that transforms according to (1.8) provides a covariant derivative and allows defining a covariant derivative and curvature tensor analogous to $R^\alpha_{\mu\nu\beta}$,

$$\left(\hat{\nabla}_\mu\right)^\beta_\alpha = \delta^\beta_\alpha \partial_\mu + \hat{\Gamma}^\beta_{\mu\alpha}, \quad \hat{R}^\alpha_{\mu\nu\beta} = [\hat{\nabla}_\mu, \hat{\nabla}_\nu]^\alpha_\beta. \tag{1.11}$$

It is not difficult to show that if the covariant derivative is required to be metric-compatible, $\nabla_\mu g_{\alpha\beta} = 0$. Then the symmetric part of the connection, $(\hat{\Gamma}^\alpha_{\mu\nu} + \hat{\Gamma}^\alpha_{\mu\nu})/2$, is necessarily given by the Christoffel symbols (1.7) [21].

Following this observation, Weyl proposed adding an extra piece, involving a vector field v_μ, to the standard metric connection [19],

$$\hat{\Gamma}^\alpha_{\mu\nu} = \Gamma^\alpha_{\mu\nu} + \frac{1}{2} g^{\alpha\sigma} \left(g_{\mu\sigma} v_\nu + g_{\sigma\nu} v_\mu - g_{\mu\nu} v_\sigma \right). \tag{1.12}$$

As a consequence of this modification metricity is lost ($\hat{\nabla}_\mu g_{\alpha\beta} = -v_\mu g_{\alpha\beta}0$), and the parallel transport of a vector acquires a scale factor in addition to the rotation produced by the metric-compatible connection. If we denote by $U_{||}(x^1)$ the parallel transported vector U from x^0 to x^1 using the metric-compatible connection, then the parallel transport under the Weyl connection yields

$$\hat{U}_{||}(x^1) = \exp\left[\int_{x_0}^{x_1} v_\mu dx^\mu\right] U_{||}(x^1). \tag{1.13}$$

Hence, Weyl's definition of parallel transport along a closed path yields the same answer as the standard Riemannian case if and only if the vector v_μ is a gradient, otherwise it would yield a path-dependent scale factor.

Because of the analogy with the non-integral nature of the electromagnetic vector potential, Weyl proposed the identification $v_\mu = \frac{e}{\gamma}A_\mu$, where γ is real constant. In this way, the Weyl connection (1.12) defines an affine structure that could unify gravitation with the electromagnetism. As noted by Einstein, however, it would mean that measurements could dependent on their histories which, according to the conventional assumptions, is not Physically acceptable. In spite of this failure, Weyl's construction was taken up by F. London [22], who proposed taking $\gamma = -i$, replacing the global scale factor by a phase rotation

$$\exp\left[i\,e\int A_\mu dx^\mu\right].$$

London applied this idea to the wave function, providing a simple way to include the electromagnetic interaction in quantum mechanics, in perfect analogy with the minimal coupling trick $\partial_\mu \to \partial_\mu - ieA_\mu$, and that was the starting point of gauge principle ideas in their simplest form, where the Lie group is identified with the abelian $U(1)$ or $SO(2)$ groups. This idea had been independently developed by Fock [23], who also understood the relevance of gauge invariance for the form of the electromagnetic coupling in the Schrödinger equation. See e.g., [24] for a complete historical discussion.

In summary, gauge symmetry introduces the minimal substitution of ordinary derivatives by covariant ones and this provides a simple universal recipe to describe interactions with gauge fields. This recipe does not require the introduction of dimensionful constants in the action, which is a welcome feature in perturbation theory since the expansion is likely to be well behaved. Also, gauge symmetry severely restricts the type of counterterms that can be added to the action, as there is a limited number of gauge-invariant expressions in a given number of spacetime dimensions. Hence, if a Lagrangian contains all possible terms allowed by the symmetry, perturbative corrections could only lead to rescalings of the coefficients in front of each term in the Lagrangian. These rescalings, in turn can be absorbed in a redefinition of the parameters of the action, which is why the renormalization procedure works in gauge systems and it is the key to the internal consistency of these theories.

1.3 Gauge Symmetry and General Coordinate Transformations

It is quite obvious that transformations (1.8) have the same form as a nonabelian gauge transformation,

$$\mathbf{A}(\mathbf{x}) \rightarrow \mathbf{A}'(\mathbf{x}) = \mathbf{U}(x)\mathbf{A}(\mathbf{x})\mathbf{U}(x)^{-1} + \mathbf{U}(x)d\mathbf{U}^{-1}(x), \tag{1.14}$$

where the "vector potential" $(A_\mu)^\alpha_\beta$ is represented by the Levi-Civita connection $\Gamma^\alpha_{\mu\beta}$, and the position-dependent group element $\mathbf{U}^\alpha_\beta(x)$ is the Jacobian matrix $\frac{\partial y^\alpha}{\partial x^\beta}$. In this sense, gravitation could be considered as a gauge theory for the group of general coordinate transformations (GCTs).[2]

The difference between gravity and a standard gauge theory for a Yang-Mills system is aggravated by the fact that the action for gravity in four dimensions cannot be written as that of a gauge-invariant system for the diffeomorphism group. In YM theories the connection $\mathbf{A}_\mu$ is an element of a Lie algebra $\mathcal{G}$, but the algebraic properties of $\mathcal{G}$ (structure constants and similar invariants) are independent of the dynamical fields or their equations. In electroweak and strong interactions, the connection $\mathbf{A}$ is dynamical, while both the base manifold and the symmetry groups are fixed, regardless of the values of $\mathbf{A}$ or the position in spacetime. This implies that the structure constants are neither functions of the field $\mathbf{A}$, or the position x. If $G^a(x)$ are the gauge generators in a YM theory, they obey an algebra of the form

$$[G^a(x), G^b(y)] = C^{ab}_c \delta(x,y) G^c(x), \tag{1.15}$$

where the coefficients C^{ab}_c are constants; whence their name **structure constants**. In contrast, the algebra of diffeomorphisms takes the form [25, 26]

$$[\mathcal{H}_\perp(x), \mathcal{H}_\perp(y)] = g^{ij}(x)\delta(x,y)_{,i}\,\mathcal{H}_j(y) - g^{ij}(y)\delta(y,x)_{,i}\,\mathcal{H}_j(x)$$

$$[\mathcal{H}_i(x), \mathcal{H}_j(y)] = \delta(x,y)_{,i}\,\mathcal{H}_j(y) - \delta(x,y)_{,j}\,\mathcal{H}_i(y)$$

$$[\mathcal{H}_\perp(x), \mathcal{H}_i(y)] = \delta(x,y)_{,i}\,\mathcal{H}_\perp(y), \tag{1.16}$$

where the $\mathcal{H}_\perp(x)$, $\mathcal{H}_i(x)$ are the generators of time and space translations, respectively, also known as the Hamiltonian constraints of gravity, and $\delta(y,x)_{,i} = \frac{\partial\delta(y,x)}{\partial x^i}$ [27]. Clearly, these generators do not form a Lie algebra but an **open algebra**, which has *structure functions* instead of *structure constants* [28]. Moreover, those coefficients playing the role of the structure constants C^{ab}_c, are also functions of the inverse metric $g^{ij}(x)$, which is a dynamical field. This means that the symmetry group in this supposed gauge theory depends on the state of the system. The

---

[2]General coordinate transformations are differentiable mappings or *coordinate diffeomorphisms*, and the set of all such transformations is often referred to as the "diffeomorphism group". Therefore, the invariance of gravity under GCTs is also interpreted as a gauge symmetry for the diffeomorphism group. As we argue next, this notion is inaccurate and misleading. Fortunately, it is also unnecessary.

structure "constants" may change from one point to another and with the dynamical evolution of the geometry, which means that the symmetry group is not uniformly defined throughout spacetime. This invalidates an interpretation of gravity in terms of fiber bundles, in which the base is spacetime and the symmetry group is the fiber. We will return to this problem in the next chapter.

Chapter 2

Geometry: General Overview

We would like to discuss now what we mean by spacetime geometry. Geometry is sometimes understood as the set of assertions that can be made about points, lines and higher-dimensional submanifolds usually embedded in a given manifold. The epitome of such logical construction is Euclid's mathematical compendium of the classical Greek period [29]. This broad (and vague) idea, is often viewed as encoded in the metric tensor, $g_{\mu\nu}(x)$, which provides the notion of distance between nearby points with coordinates x^{μ} and $x^{\mu} + dx^{\mu}$,

$$ds^2 = g_{\mu\nu} \, dx^{\mu} dx^{\nu}. \tag{2.1}$$

This is the case in Riemannian geometry, where all relevant objects defined on the manifold (distance, area, angles, parallel transport operations, curvature, etc.) can be constructed from the metric. However, a distinction should be made between *metric* and *affine* features of spacetime.

Metricity refers to measurements of lengths, angles, areas or volumes of objects, which are locally defined in spacetime. **Affinity** refers to properties which remain invariant under translations — or more generally, affine transformations — such as parallelism.

The distinction is useful because these two notions are logically independent, and reducing one to the other is an unnecessary form of violence.

2.1 Metric and Affine Structures

Euclidean geometry was constructed using two elementary instruments: the compass and the unmarked straightedge. The first is a metric instrument because it allows comparing lengths and, in particular, drawing circles. The second is used to draw straight lines which, as will be seen below, is a basic affine operation.

Pythagoras' famous theorem is a metric statement; it relates the lengths of the sides of a triangle in which two sides form a particular angle. Affine properties on the other hand, do not change if the scale is changed, as for example the shape of a triangle or the angle between two straight lines. A typical affine statement is, for instance, the fact that when two parallel lines intersect a third, the corresponding angles are equal.

Of course parallelism can be reduced to metricity. As we all learned in school, one can draw a parallel to a line using a compass and an unmarked straightedge: One draws two circles of equal radii with centers at two points of one line and then draws the tangent to the two circles with the ruler. Thus, given a way to measure distances and straight lines in space, one can define parallel transport.

As any child knows from the experience of stretching a string or a piece of rubber band, a straight line is the shape that minimizes the distance between two points. This definition of a straight line is clearly a metric statement because it requires *measuring* lengths. Orthogonality is also a metric notion that can be defined using the scalar product obtained from the metric. A right angle is a metric feature because in order to build one, one should be able to *measure* angles, or measure the sides of triangles[1]. We will now argue that *parallelism does not require metricity.*

There is something excessive about using a compass for the construction of a parallel to a straight line through a given point. In fact, a rigid wedge of any fixed angle would suffice: align one of the sides of the wedge with the straight line and rest the straightedge on other side; then slide the wedge along the straightedge to reach the desired point. The only requirement for the wedge is *not to change* in the process so that the angle between its sides should be preserved. Thus, the essence of parallel transport is a rigid wedge and a straightedge to connect two points, but no measurements of lengths or angles need to be involved.

There is still some cheating in this argument because we took the construction of a straightedge for granted. How do we construct a straight line if we had no notion of distance? After all, as we discussed above, a straight line is often defined as the shortest path between two points. Fortunately, there is a purely affine way to construct a straight line: Take two short enough segments (two short sticks, matches or pencils would do), and slide them one along the other, as a cross country skier would do. In this way a straight line is generated by *parallel transport of a vector along itself,* and we have not used distance anywhere. It is essentially this *affine* definition of a straight line that can be found in Book I of Euclid's *Elements.* This definition could be regarded as the *straightest line,* which need not coincide with the *line of shortest distance.* In fact, if the two sticks one uses are two identical arcs, one could construct families of "straight" lines by parallel transport of a segment along

---

[1] The Egyptians knew how to *use* Pythagoras' theorem to make a right angle, although they didn't know how to prove the theorem (and probably never worried about it). Their recipe was probably known long before, and all good construction workers today still know it: make a loop of rope with 12 segments of equal length. Then, the triangle formed with the loop so that its sides are 3, 4 and 5 segments long is such that the shorter segments are perpendicular to each other [30].

itself, but they would not correspond to shortest lines on a plane. They are logically independent concepts. The purely affine construction is logically acceptable, which means that parallel transport is not necessarily a metric operation.

In a space devoid of a metric structure the straightest line could be a rather weird looking curve, but it could still be used to define parallelism. If a ruler has been constructed by transporting a vector along itself, then one can use it to define parallel transport, completely oblivious that the straight lines are not necessarily the shortest. There would be nothing wrong with such a construction except that it will probably not coincide with the standard metric construction.

A general feature of any local definition of parallelism is that it cannot be consistently defined in an open region of an arbitrarily curved manifold. The problem is that in general, parallelism is not a transitive relation if parallel transport is performed along different curves. Take for example an ordinary two-dimensional sphere. The construction of a "straight" line by parallel transport of a small segment along itself yields a great circle, defined as the intersection of the sphere with a plane that passes through the center of the sphere. Consider a triangle formed by three segments of great circles. It is possible to transport a vector around the triangle in such a way that the angle between the vector and the arc is kept constant, as outlined above. Then, it is easy to see that the vector obtained by parallel transport around the three sides of the triangle does not coincide with the original vector. This discrepancy between a vector and its parallel-transport along a closed loop is a clear violation of the transitivity property: if $\vec{a}$ is parallel to $\vec{b}$, and $\vec{b}$ is parallel to $\vec{c}$, it does not follow that $\vec{a}$ is parallel to $\vec{c}$. Therefore, on an open neighborhood of a curved manifold, the notion of parallelism need not be an equivalence relation.

Here the problem arises when parallel transport is performed along different curves, so one could expect to see no conflict for parallelism along a single "straight line". However, this is also false as can be checked by parallel transporting a vector along a straight line that encloses the apex of a cone[2] with angular deficit $\Delta > \pi$. In this case, the problem stems from trying to define parallel transport on a region that encloses a singularity of infinite curvature, as will be discussed in Section 6.2, on naked singularities and branes.

Note that the effect of curvature only depends on the definition of parallel transport along a curve, which is also a purely affine notion. This is a general feature: the curvature of a manifold is completely determined by the definition of parallel transport, independently of the metric.

---

[2]Such a cone can be made by identifying the two radii of a flat angular sector. If the central angle of the sector is θ, the cone has angular deficit $\Delta = 2\pi - \theta$. If $\theta < \pi$ (and consequently $\Delta > \pi$) it is possible to draw a line that intersects the two radii. When the two radii are identified this straight line self-intersects and winds around the apex of the cone.

2.1.1 *Metric structure*

Metricity is encoded in a rule to define distance between neighboring points in a manifold. It is a differential notion of distance which can be integrated to define the length of a path or the area enclosed by a curve, or more generally the volume of a region in a manifold. The fundamental expression is the quadratic form

$$ds^2 = g_{\mu\nu}(x)dx^\mu dx^\nu \,, \tag{2.2}$$

which expresses the length of an infinitesimal segment between x and $x+dx$ squared. The metric $g_{\mu\nu}(x)$ is supposed to be nonsingular (invertible) throughout the manifold, with the possible exception of sets of measure zero (singularities). Under a change of coordinates, $x \to \tilde{x}(x)$,

$$(d\tilde{s})^2 = \tilde{g}_{\mu\nu}(\tilde{x})d\tilde{x}^\mu d\tilde{x}^\nu$$
$$= \tilde{g}_{\mu\nu}(\tilde{x})\frac{\partial \tilde{x}^\mu}{\partial x^\alpha}dx^\alpha \frac{\partial \tilde{x}^\nu}{\partial x^\beta}dx^\beta \,. \tag{2.3}$$

Hence, if the notion of distance is to have an invariant meaning, irrespective of the coordinates used, the metric must transform as a (covariant) tensor under general coordinate transformations,

$$\tilde{g}_{\mu\nu}(\tilde{x}) = \frac{\partial x^\alpha}{\partial \tilde{x}^\mu} \frac{\partial x^\beta}{\partial \tilde{x}^\nu} g_{\alpha\beta}(x). \tag{2.4}$$

Notice that the metric, as any nondegenerate quadratic form, defines a scalar product for vectors and a norm in the manifold. This is very important in physics in order to define invariant observables like the mass, the proper time, etc. The fact that the spacetime is locally Lorentzian requires that in the neighborhood of any point in the manifold there always exists a coordinate patch such that the metric takes the form $g_{\mu\nu}(x) = \text{diag}(-1, 1, 1, \cdots, 1)$.

2.1.2 *Connection*

As already mentioned, parallelism is encoded in an affine structure known as a *connection*. Consider a vector field with components $u^r(x)$ at the point with coordinates x^μ. The vector $u^r_{||}(x)$ denotes the parallel-transported of the vector $u^r(x+dx)$ from $x+dx$ to x if their components are related by

$$u^r_{||}(x) = u^r(x+dx) + dx^\mu \Theta^r_{s\mu}u^s(x)$$
$$= u^r(x) + dx^\mu[\partial_\mu u^r + \Theta^r_{s\mu}u^s(x)]. \tag{2.5}$$

The coefficients $\Theta^r_{s\mu}(x)$ define parallelism and this is related to the group under which $u^r(x)$ transforms as a vector. The important feature of a connection is that it can be used to define the covariant derivative, a differential operation that takes a vector into a vector. This is possible because $u^\alpha_{||}(x)$ and $u^\alpha(x)$ are two vectors at the same point and therefore their difference,

$$u_{\parallel}^r(x) - u^r(x) = dx^\mu[\partial_\mu u^r + \Theta_{s\mu}^r u^s(x)]$$

$$\equiv dx^\mu[\mathbf{D}_\mu u(x)]^r, \tag{2.6}$$

transforms homogeneously under the local action of the group for which u^r is a vector. In order to achieve this, the connection itself should transform in a precise way as in (1.8) or (1.14). The word "parallel" is a metaphor that evokes the idea borrowed from geometry of drawing a parallel to a line through a point elsewhere. Parallelism here refers to a general rule that allows to define a symmetric reflective relation between vectors infinitesimally separated. In Riemannian geometry the definition of parallelism is provided by the Levi-Civita form (Γ) and is related to the group of general coordinate transformations, the diffeomorphism group; in nonabelian gauge theories the relation involves the Lie-algebra valued gauge potential ($\mathbf{A}$).

As already mentioned, the affine connection $\Gamma_{\mu\beta}^\alpha(x)$ need not be functionally related to the metric tensor $g_{\mu\nu}(x)$. However, Einstein formulated General Relativity adopting the point of view that the spacetime metric should be the only dynamically independent field, while the affine connection should be a function of the metric given by the Christoffel symbol (1.7),

$$\Theta_{\mu\beta}^\alpha =: \Gamma_{\mu\beta}^\alpha = \frac{1}{2}g^{\alpha\lambda}(\partial_\mu g_{\lambda\beta} + \partial_\beta g_{\lambda\mu} - \partial_\lambda g_{\mu\beta}). \tag{2.7}$$

This was the starting point of a controversy between Albert Einstein and Elie Cartan, which is vividly recorded in the correspondence they exchanged between May 1929 and May 1932 [31]. In his letters, Cartan politely but forcefully insisted that metricity and parallelism could be considered as independent, while Einstein pragmatically replied that since the space we live in seems to have a metric, it would be more economical to assume the affine connection to be a function of the metric. Cartan advocated economy of assumptions. Einstein argued in favor of economy of independent fields.

Here we adopt Cartan's point of view. It is less economical in dynamical variables but more economical in assumptions and therefore more general. This alone would not be sufficient ground to adopt Cartan's philosophy, but it turns out to be more transparent in many ways, and lends itself better to discuss the differences and similarities between gauge theories and gravity. Moreover, Cartan's point of view emphasizes the distinction between the spacetime manifold M as the base of a fiber bundle, and the tangent space at every point T_x, where Lorentz vectors, tensors and spinors live.

In physics, the metric and the connection play different roles as well. The metric enters in the definition of the kinetic energy of particles and of the stress-energy tensor density in a field theory, $T^{\mu\nu}$. The spin connection instead, appears in the coupling of fermionic fields with the geometry of spacetime. All forms of matter are made out of fermions (leptons and quarks), so if we hope someday to attain a unified description of all interactions including gravity and matter, it is advisable to grant the affine structure of spacetime an important role, on equal footing with the metric structure.

2.2   Nonabelian Gauge Fields

The logical independence between a metric and the affine structure is seen most clearly in nonabelian gauge theory, where the connection is the "vector potential" $\Theta^r_{s\mu} = (\mathbf{A}_\mu)^r_s$, defined solely by the condition that under a gauge transformation, it changes as in (1.14),

$$\mathbf{A}_\mu(x) \rightarrow \mathbf{A}'_\mu(x) = \mathbf{U}(x)\mathbf{A}_\mu(x)\mathbf{U}(x)^{-1} + \mathbf{U}(x)\partial_\mu\mathbf{U}^{-1}(x). \qquad (2.8)$$

Here $\mathbf{U}(x)$ represents a position-dependent group element and there is no reference to a metric. In general there is no naturally defined notion of metric structure in a Lie group, and even if the connection field were defined in a manifold endowed metric $g_{\mu\nu}$, that is irrelevant for (2.8). The value of $\mathbf{A}$ depends on the choice of $\mathbf{U}(x)$, and $\mathbf{A}$ can even be made to vanish at a given point by an appropriate gauge transformation.

The covariant derivative

$$\mathbf{D}_\mu = \partial_\mu + \mathbf{A}_\mu\,, \qquad (2.9)$$

is a differential operator that, unlike the ordinary derivative and $\mathbf{A}$ itself, transforms homogeneously under the action of the gauge group,

$$\mathbf{D}_\mu \rightarrow \mathbf{D}'_\mu = \mathbf{U}(x)\mathbf{D}_\mu\mathbf{U}(x)^{-1}. \qquad (2.10)$$

The operator $\mathbf{D}_\mu$ is in general a matrix-valued 1-form,

$$\begin{aligned} \mathbf{D} &= d + \mathbf{A} \\ &= dx^\mu(\partial_\mu + \mathbf{A}_\mu). \end{aligned} \qquad (2.11)$$

If $\phi(x)$ is a function in a vector representation of the gauge group ($\phi(x) \rightarrow \phi'(x) = \mathbf{U}(x) \cdot \phi(x)$), its covariant derivative reads

$$\mathbf{D}\phi = d\phi + \mathbf{A} \wedge \phi. \qquad (2.12)$$

As was seen in the case of the Levi-Civita connection (1.10), the covariant derivative operator $\mathbf{D}$ in (2.12) has the remarkable property that its square is not a differential operator but a multiplicative one,

$$\begin{aligned} \mathbf{DD}\phi &= d(\mathbf{A}\phi) + \mathbf{A}d\phi + \mathbf{A} \wedge \mathbf{A}\phi \\ &= (d\mathbf{A} + \mathbf{A} \wedge \mathbf{A})\phi \\ &= \mathbf{F}\phi. \end{aligned} \qquad (2.13)$$

The combination $\mathbf{F} = d\mathbf{A} + \mathbf{A} \wedge \mathbf{A}$ is the curvature two-form, also known in physics as field strength of the non-abelian interaction, that generalizes the electric and magnetic fields of electromagnetism.

One can now see why the gauge principle is such a powerful idea in physics: the covariant derivative of a field, $\mathbf{D}\phi$, defines the coupling between ϕ and the gauge potential $\mathbf{A}$ in a unique way, and this coupling is by construction gauge invariant. Moreover, $\mathbf{A}$ has a uniquely defined field strength $\mathbf{F}$, which in turn defines the dynamical properties of the gauge field. In 1954, Robert Mills and Chen-Nin Yang grasped the beauty and power of this idea and constructed what has been known as the non-abelian Yang-Mills theory [7], which is the fundamental theory behind the Standard Model of subatomic physics.

In summary, when dealing with a nonabelian gauge interaction in a curved manifold, there are two covariant derivative operators, the differential geometry one of the metric spacetime manifold,

$$\nabla = d + \mathbf{\Gamma}$$
$$= dx^\mu (\partial_\mu + \mathbf{\Gamma}_\mu), \qquad (2.14)$$

and the one of the internal gauge symmetry (2.11). If the components of the connection $\mathbf{\Gamma}$ are chosen as the Christoffel connection (1.7), the transformations properties of $\mathbf{\Gamma}$ under general coordinate transformations are (1.8), and the differential operator ∇ is covariant under the diffeomorphism group. To that extent $\mathbf{\Gamma}$ acts as a connection; however this is not enough to turn gravity into a gauge theory. The problem is that this group acts on the coordinates of the manifold as $x^\mu \to x'^\mu(x) = x^\mu + \xi^\mu(x)$, which is a shift in the arguments of the fields (tensors) on which it acts. On the other hand, a gauge transformation in the sense of fiber bundles, acts on the functions and not on their arguments, i.e., it generates a motion along the fiber at a fixed point on the base manifold. For this reason, $\mathbf{\Gamma}$ is not a gauge connection in the same footing with the nonabelian connection $\mathbf{A}$.

Chapter 3

First Order Gravitation Theory

On November 25 1915, Albert Einstein presented to the Prussian Academy of Natural Sciences the equations for the gravitational field in the form we now know as Einstein equations [32]. Curiously, five days before, David Hilbert had proposed the correct action principle for gravity, based on a communication in which Einstein had outlined the general idea of what should be the form of the equations [33]. As we shall see, this is not so surprising in retrospect, because there is a unique action which is compatible with the postulates of general relativity in four dimensions that admits flat space as a solution. If one allows constant curvature geometries, there is essentially a one-parameter family of actions that can be constructed: the Einstein-Hilbert form plus a cosmological constant term,

$$I[g] = \int \sqrt{-g}(\alpha_1 R + \alpha_0)d^4x, \qquad (3.1)$$

where the scalar curvature R is a function of the metric $g_{\mu\nu}$, its inverse $g^{\mu\nu}$, and its derivatives (we follow the definitions and conventions of Misner, Thorne and Wheeler [34]). The action $I[g]$ is the only functional in four dimensions which is invariant under general coordinate transformations and gives second order field equations for the metric. The coefficients α_1 and α_0 are related to the Newton's constant and the cosmological constant through

$$G = \frac{1}{16\pi\alpha_1} \ , \ \Lambda = -\frac{\alpha_0}{2\alpha_1}. \qquad (3.2)$$

The Einstein equations are obtained by extremizing the action (3.1) with respect to $g^{\mu\nu}$,

$$G_{\mu\nu} + \Lambda g_{\mu\nu} = 0, \qquad (3.3)$$

where $G_{\mu\nu}$ is the Einstein tensor defined by

$$G_{\mu\nu} = R_{\mu\nu} - \frac{R}{2}g_{\mu\nu}. \qquad (3.4)$$

These equations are unique in the sense that:

 (i) They are tensorial.
 (ii) They involve only up to second derivatives of the metric.
 (iii) They reproduce Newtonian gravity in the non-relativistic weak field approximation.

The first condition implies that the equations have the same meaning in all coordinate systems. This follows from the need to have a coordinate-independent (covariant) formulation of gravity in which the gravitational force is replaced by the non-flat geometry of spacetime. The gravitational field is a geometrical entity and therefore its existence cannot depend on the coordinate choice or, in physical terms, the choice of reference frame. Of course, the *expression* of the gravitational interaction in different reference frames need not be the same, but the changes should not be arbitrary. On the contrary, under a change of coordinates $x \to x'(x)$ the expression of the field that describes the gravitational interaction should transform in a *representation* of the group of coordinate transformations. This is what is meant by saying that the metric $g_{\mu\nu}$, as well as the equations that determine it, transform as tensors.

Condition (ii) means that Cauchy data are necessary (and sufficient in most cases) to integrate the equations. This condition is a concession to the classical physics tradition: the possibility of determining the gravitational field at any moment from the knowledge of the configuration of space, $g_{ij}(x)$, and its time derivative $\dot{g}_{ij}(x)$, at a given time. This requirement is also the hallmark of Hamiltonian dynamics, which is the starting point for canonical quantum mechanics and therefore suggests that a quantum version of the theory could exist. Covariance under general spacetime coordinate transformations requires the equations to be second order in spatial derivatives as well.

The third requirement is the correspondence principle, which accounts for our daily experience that an apple and the moon fall the way they do. In other words, General Relativity should include Newtonian gravitation in the static, weak field limit.

If in addition one assumes that Minkowski space be among the solutions of the matter-free theory, then one must set $\Lambda = 0$, as most sensible particle physicists would do. If, on the other hand, one believes in static homogeneous and isotropic cosmologies, as Einstein did, then Λ must have a finely tuned nonzero value. Experimentally, Λ has a value of the order of 10^{-120} in "natural" units ($\hbar = c = G = 1$) [21]. Furthermore, astrophysical measurements seem to indicate that Λ must be positive [35]. This presents a problem because there seems to be no theoretical way to predict this "unnaturally small" positive value.

This summarizes the so-called metric formulation of gravity, in which both the metric and affine features of the spacetime geometry are determined by the metric field $g_{\mu\nu}$. In what follows, two fields will be used to characterize the geometry: the vielbein and the Lorentz connection are introduced as the fundamental dynamical objects codifying the metric and affine features, respectively. It will be shown how the Einstein–Hilbert action (3.1) can be constructed using only these ingredients, their exterior derivatives (torsion and curvature) and some appropriate Lorentz invariant tensors. As a direct consequence, pure gravity in four dimensions can be shown to be a genuine gauge theory of the local Lorentz symmetry group. As we will see later, for dimensions $d > 4$, the Einstein–Hilbert action is not the only Lorentz gauge invariant theory that satisfies the conditions (**i**)-(**iii**) and their number grows wildly with the number of spacetime dimensions.

3.1 The Equivalence Principle

As early as 1907, Einstein observed that the effect of gravity can be neutralized by free fall: In a freely falling laboratory, one cannot feel gravity and any released object would float in front of our eyes, as experienced by the crew of a spaceship, precisely because the ship is in a freely falling orbit. This trick is a local one: the lab has to be *small enough* and the time span of the experiments must be *short enough*. Under these conditions, the experiments will be indistinguishable from those performed in absence of gravity, and the laws of physics that will be reflected by the experiments will be those valid in Minkowski space, i.e., Lorentz invariance. This means that, in a local neighborhood, spacetime possesses Lorentz invariance. In order to make manifest this invariance, it is necessary to perform an appropriate coordinate transformation to a particular reference system, viz., a freely falling one.

Conversely, Einstein argued that in the absence of gravity, the gravitational field could be mocked by applying an acceleration to the laboratory. This idea is known as the principle of equivalence, meaning that gravitation and acceleration are equivalent effects in a small spacetime region.

A freely falling observer defines a *locally inertial system*. For a small enough region around the freely falling observer the trajectories of projectiles (freely falling as well) are straight lines and the discrepancies with Euclidean geometry are negligible. Particle collisions mediated by short range forces, such as those between billiard balls, molecules or subnuclear particles, satisfy the conservation laws of energy and momentum valid in special relativity.

How large is the neighborhood where this approximation is good enough? Well, it depends on the accuracy one wants to achieve, and on how curved spacetime really is in that neighborhood. In the region between the earth and the moon, for instance, the deviation from flat geometry is about one part in 10^9. Whether this is good enough or not, it depends on the experimental accuracy of the tests involved (see Ref. [36]), but whatever the accuracy one wishes to achieve, there is always a small enough neighborhood where the curved spacetime is accurately described by a flat one. In differential geometry, this flat space corresponds to the tangent space that can be defined at every point of a smooth manifold.

Thus, we view spacetime as a smooth D-dimensional manifold M. At every point $x \in M$ there exists a D-dimensional flat tangent space T_x of Lorentzian signature $(-, +, \ldots, +)$. This tangent space T_x is a good approximation of the manifold M on an open set in the neighborhood of x. In fact, this is the defining property of a differentiable manifold of dimension D, one that at each point has an open neighborhood isomorphic to a patch in $\mathbb{R}^D$. The transition between a neighborhood in M and a neighborhood in T_x is a change of reference frame to that of a freely falling observer, which is just a change in the spacetime coordinates. This means that there is a way to represent tensors on M by tensors on T_x, and vice versa. The precise relation between the tensor spaces on M and on T_x is an isomorphism represented by means of a linear map e called the *vielbein*.

3.1.1   *The vielbein*

The isomorphism between M and the collection of tangent spaces at each point of the manifold $\{T_x\}$ can be explicitly constructed as a coordinate transformation between a system of local coordinates $\{x^\mu\}$ in an open neighborhood in M and an orthonormal frame in the Minkowski space T_x with coordinates z^a. The Jacobian matrix

$$\frac{\partial z^a}{\partial x^\mu} = e^a_\mu(x) \tag{3.5}$$

is sufficient to define a one-to-one relation between tensors in each space: if Π is a tensor with components $\Pi^{\mu_1 \cdots \mu_n}(x)$ in M, then the corresponding tensor in the tangent space T_x is[1]

$$P^{a_1 \cdots a_n}(x) = e^{a_1}_{\mu_1}(x) \cdots e^{a_n}_{\mu_n}(x) \Pi^{\mu_1 \cdots \mu_n}(x). \tag{3.6}$$

In particular, since Minkowski space has a canonically defined metric η_{ab}, this isomorphism induces a metric on M. Consider the coordinate separation dx^μ between two infinitesimally close points on M. The corresponding separation dz^a in T_x is

$$dz^a = e^a_\mu(x) dx^\mu. \tag{3.7}$$

Then, the arc length in T_x, $ds^2 = \eta_{ab} dz^a dz^b$, can also be expressed as $\eta_{ab} e^a_\mu(x) e^b_\nu(x) dx^\mu dx^\nu$, where the metric in M is identified,

$$g_{\mu\nu}(x) = e^a_\mu(x) e^b_\nu(x) \eta_{ab}. \tag{3.8}$$

This relation can be interpreted as the vielbein being the "square root" of the metric. Since $e^a_\mu(x)$ determines the metric, all metric properties of spacetime are contained in the vielbein. The converse, however, is not true: given a metric $g_{\mu\nu}(x)$, there exist infinitely many choices of vielbein that reproduce the same metric. This infinitely

---

[1]The inverse vielbein, $e^\mu_a(x)$, where $e^\nu_a(x) e^b_\nu(x) = \delta^b_a$, and $e^\nu_a(x) e^a_\mu(x) = \delta^\nu_\mu$, relates lower index tensors, $P_{a_1 \ldots a_n}(x) = e^{\mu_1}_{a_1}(x) \cdots e^{\mu_n}_{a_n}(x) \Pi_{\mu_1 \ldots \mu_n}(x)$.

many choices of vielbein correspond to the different choice of local orthonormal frames that can be used as bases for the tangent space vectors at T_x. It is possible to rotate the vielbein by a Lorentz transformation and this should be undetectable from the point of view of the manifold M.

Under a Lorentz transformation, the vielbein transforms as a vector

$$e_\mu^a(x) \longrightarrow e_\mu'^a(x) = \Lambda^a_b(x)e_\mu^b(x), \tag{3.9}$$

where the matrix $\Lambda(x) \in SO(D-1,1)$. By definition of $SO(D-1,1)$, $\Lambda(x)$ leaves the metric in the tangent space unchanged,

$$\Lambda^a_c(x)\Lambda^b_d(x)\eta_{ab} = \eta_{cd}. \tag{3.10}$$

The metric $g_{\mu\nu}(x)$ is clearly also unchanged by this transformation. This means, in particular, that there are many more components in e_μ^a than in $g_{\mu\nu}$. In fact, the vielbein has D^2 independent components, whereas the metric has only $D(D+1)/2$. The mismatch is exactly $D(D-1)/2$ which corresponds to the number of independent rotations in D dimensions.

Thus, the metric structure of the manifold is contained in a local orthonormal frame $\{e_\mu^a(x), a = 1, \ldots, D = dimM\}$, also called *"soldering form"*, *"moving frame"*, or simply, **vielbein**. The definition (3.5) implies that $e_\mu^a(x)$ transforms as a covariant vector under diffeomorphisms on M and as a contravariant vector under local Lorentz rotations of T_x, $SO(D-1,1)$ (the signature of the manifold M is assumed to be Lorentzian).

3.1.2 *The Lorentz connection*

The invariance of the tangent space under the action of the Lorentz group at each point of M endows the manifold with a gauge symmetry, the invariance under local Lorentz rotations. Mathematically, this is captured by the concept of fiber bundle: a family of vector spaces $\{T_x\}$ labeled by the points of M, the tangent bundle. As discussed in the preceding chapters, in order to define a covariant derivative operator on the manifold, a connection is required so that the differential structure remains invariant under local Lorentz transformations acting independently at each spacetime point. This connection, denoted by $\omega^a_{b\mu}$, is usually called "spin connection" in the physics literature, although *Lorentz connection* is a more appropriate name. The word "spin" is due to the fact that $\omega^a_{b\mu}$ arises naturally in the discussion of spinors, which carry a special representation of the group of rotations in the tangent space, but that is irrelevant at the moment.[2]

If ϕ is a tensor in T_x, its covariant derivative $D\phi$ is a tensor of the same rank and nature as ϕ. In order to do this, a connection is needed to compensate for the fact that the gauge group acts independently at neighboring points x and $x + dx$.

---

[2]Here, only the essential ingredients are given. For a more extended discussion, there are several texts such as those of Refs. [8], [9], [37], [38] and [39].

Suppose $\phi^a(x)$ is a field that transforms as a vector under the Lorentz group, its covariant derivative,

$$D_\mu \phi^a(x) = \partial_\mu \phi^a(x) + \omega^a{}_{b\mu}(x)\phi^b(x), \qquad (3.11)$$

must also transforms as a Lorentz vector at x. This requires that under an $SO(D-1,1)$ rotation, $\Lambda^a{}_c(x)$, the connection changes as (1.14), which in this case reads

$$\omega^a{}_{b\mu}(x) \to \omega'^a{}_{b\mu}(x) = \Lambda^a{}_c(x)\Lambda_b{}^d(x)\omega^c{}_{d\mu}(x) + \Lambda^a{}_c(x)\partial_\mu\Lambda_b{}^c(x), \qquad (3.12)$$

where $\Lambda_b{}^d = \eta_{ab}\eta^{cd}\Lambda^a{}_c$ is the inverse (transpose) of $\Lambda^b{}_d$.

The connection $\omega^a{}_{b\mu}(x)$ defines the *parallel transport* of Lorentz tensors in the tangent space between T_x and T_{x+dx}. The parallel transport of the vector field $\phi^a(x)$ from $x + dx$ to x, is a vector $\phi^a_{||}(x)$, defined as

$$\phi^a_{||}(x) \equiv \phi^a(x+dx) + dx^\mu \omega^a{}_{b\mu}(x)\phi^b(x), \qquad (3.13)$$

$$= \phi^a(x) + dx^\mu[\partial_\mu\phi^a(x) + \omega^a{}_{b\mu}(x)\phi^b(x)],$$

$$:= \phi^a(x) + dx^\mu D_\mu\phi^a(x).$$

Thus, the covariant derivative D_μ measures the change in a tensor produced by parallel transport between two neighboring points,

$$dx^\mu D_\mu\phi^a(x) = \phi^a_{||}(x) - \phi^a(x) \qquad (3.14)$$

$$= dx^\mu[\partial_\mu\phi^a + \omega^a{}_{b\mu}(x)\phi^b(x)].$$

In this way, the affine properties of space are encoded in the components $\omega^a{}_{b\mu}(x)$, which are, until further notice, totally arbitrary and independent from the metric. Note the similarity between the notion of parallelism in (3.14) and that for vectors whose components are referred to a coordinate basis (2.6). These two definitions are independent as they refer to objects on different spaces, but they can be related using the local isomorphism between the base manifold and the tangent space, provided by e^a_μ.

3.1.3 *A tale of two covariant derivatives*

In the previous chapters we dealt with the covariant derivative of the base manifold ∇_μ, (1.6) that takes a tensor under diffeomorphisms into another tensor with one additional (covariant) index. The Lorentz covariant operator $\mathbf{D}_\mu$ performs a similar task and since tensors in the tangent space can be translated into tensors in the base and vice-versa using the vielbein (3.6), one might expect the two derivative to be related. Such relation should involve the Lorentz connection ω, the Christoffel symbol $\Gamma^\nu_{\mu\lambda}$ and the vielbein e^a_μ.

Let us compare the action of D_μ and ∇_μ on an arbitrary vector field X. If X^μ transforms as a vector under diffeomorphisms, its covariant derivative according to (1.6) is

$$\nabla_\mu X^\nu = \partial_\mu X^\nu + \Gamma^\nu_{\mu\lambda}X^\lambda. \qquad (3.15)$$

On the other hand, X also defines a vector in the tangent space with components

$$X^a = e^a_\lambda X^\lambda, \tag{3.16}$$

and therefore, one can act on it with the Lorentz covariant derivative as defined by (3.15),

$$\mathbf{D}_\mu X^a = \partial_\mu X^a + \omega^a{}_{b\mu} X^b. \tag{3.17}$$

Now, there is problem because ∇ and $\mathbf{D}$ have well-defined actions on tensors under diffeomorphisms and under the Lorentz group, respectively. However, expressions like $\mathbf{D}_\mu X^\lambda$ or $\nabla_\mu X^a$ are not clearly defined. In particular, it would be desirable to have a well-defined notion of covariant derivative applied to a hybrid object like the vielbein e^a_μ that carries a representation of both groups. The resolution of this ambiguity calls for a generalized definition: If an object is a tensor in some tensorial representation of the group $G = G_1 \times G_2$, then the *full covariant derivative*, which we provisionally denote by $\mathfrak{D}$ —, is the differential operator that reduces to the covariant derivative of each factor when the connection in the other factor vanishes. For example, the full covariant derivative of vector under both groups $S^{a\nu}$ reads

$$\mathfrak{D}_\mu S^{a\nu} = \partial_\mu S^{a\nu} + \omega^a{}_{b\mu} S^{b\nu} + \Gamma^\nu_{\mu\lambda} S^{a\lambda}. \tag{3.18}$$

It is straightforward to check that this operator is well defined and satisfies Leibniz's rule. Applying $\mathfrak{D}_\mu$ to both sides of (3.16), gives

$$\mathfrak{D}_\mu X^a = (\mathfrak{D}_\mu e^a_\lambda) X^\lambda + e^a_\lambda \mathfrak{D}_\mu X^\lambda, \tag{3.19}$$

In the left hand side the derivative acts on a Lorentz vector and therefore $\mathfrak{D}$ can be substituted by $\mathbf{D}$. Since in the last term on the right hand side is acting on a coordinate vector, $\mathfrak{D}$ can be replaced by ∇. Combining all these terms, one finds

$$[\mathfrak{D}_\mu e^a_\lambda - \partial_\mu e^a_\lambda - \omega^a{}_{b\mu} e^b_\lambda + \Gamma^\rho_{\mu\lambda} e^a_\rho] X^\lambda = 0. \tag{3.20}$$

Since this expression should be valid for any vector X, the full covariant derivative of the vielbein must be given by

$$\mathfrak{D}_\mu e^a_\lambda = \partial_\mu e^a_\lambda + \omega^a{}_{b\mu} e^b_\lambda - \Gamma^\rho_{\mu\lambda} e^a_\rho. \tag{3.21}$$

This expression is similar to the definition of torsion in Section 3.2 and we will come back to this equation in relation in that discussion.

As discussed in Chapter 1, it is convenient to postulate that the metric should be invariant under parallel transport, which means that it must be covariantly constant,

$$\nabla_\lambda g_{\mu\nu} \equiv 0. \tag{3.22}$$

Writing the metric in terms of the vielbein, this postulate reads

$$2\eta_{ab}(\mathfrak{D}_\lambda e^a_\mu) e^b_\nu = 0. \tag{3.23}$$

Hence, assuming the vielbein to be invertible, this leads to the so-called *vielbein postulate*,

$$\partial_\mu e^a_\lambda + \omega^a{}_{b\mu} e^b_\lambda - \Gamma^\rho_{\mu\lambda} e^a_\rho = 0. \tag{3.24}$$

3.1.4 *Invariant tensors*

As will become clear later, the notion of invariant tensor is essential in the construction of Chern-Simons (super)gravity actions. An invariant tensor of rank $(n+1)$ denoted by $\Delta_{A_1 \cdots A_{n+1}}$ of a given Lie algebra (this concept also applies in the case of super Lie algebras) with generators X_A is a tensor which satisfies

$$C^F_{BA_1} \Delta_{FA_2 \cdots A_{n+1}} + \cdots + C^F_{BA_{n+1}} \Delta_{A_1 \cdots A_n F} = 0, \tag{3.25}$$

where the C^C_{AB}'s are the structure constants

$$[X_A, X_B] = C^C_{AB} X_C. \tag{3.26}$$

It is easy to see that the group $SO(D-1,1)$ has two invariant tensors, the Minkowski metric, η_{ab}, and the totally antisymmetric Levi-Civita tensor, $\epsilon_{a_1 a_2 \cdots a_D}$. These tensors are defined by the algebraic structure of the Lorentz group and therefore they are the same in every tangent space and, consequently, they must be constant throughout the manifold M, that is $d\eta_{ab} = 0 = d\epsilon_{a_1 a_2 \cdots a_D}$. Moreover, since they are invariant, they must also be covariantly constant,

$$d\eta_{ab} = D\eta_{ab} = 0, \tag{3.27}$$

$$d\epsilon_{a_1 a_2 \cdots a_D} = D\epsilon_{a_1 a_2 \cdots a_D} = 0. \tag{3.28}$$

Requiring the Lorentz connection to be compatible with the invariance of the Lorentz metric (3.27) restricts ω_{ab} to be antisymmetric,

$$\omega_{ab} = \eta_{ac} \omega^c{}_b = -\eta_{bc} \omega^c{}_a = -\omega_{ba}. \tag{3.29}$$

This is a property shared by the connection of any orthogonal group $SO(n,m)$, where the Lorentz metric is replaced by the corresponding $n+m$ invariant tensor $\hat{\eta} = \text{diag}(+ \cdots +, - \cdots -)$. On the other hand, invariance of the Levi-Civita tensor implies

$$\epsilon_{b_1 a_2 \cdots a_D} \omega^{b_1}{}_{a_1} + \epsilon_{a_1 b_2 \cdots a_D} \omega^{b_2}{}_{a_2} + \cdots + \epsilon_{a_1 a_2 \cdots b_D} \omega^{b_D}{}_{a_D} = 0, \tag{3.30}$$

which does not impose further restrictions on the components of the Lorentz connection.

The connection $\omega^a{}_{b\mu}$ has $D^2(D-1)/2$ independent components, which is *less* than the number of independent components of the Christoffel symbol, $D^2(D+1)/2$. This can be understood in the way that the Christoffel symbols are determined by both the vielbein and the Lorentz connection, $\Gamma^\lambda_{\mu\nu} = \Gamma^\lambda_{\mu\nu}(e, \omega)$. Indeed, the difference between the number of independent components of ω^{ab} and of the Christoffel symbol is D^2 which is precisely the number of independent components of the vielbein.

3.2   Curvature and Torsion Two-Forms

It can be observed that both the vielbein and the spin connection arise through the combinations

$$e^a(x) \equiv e^a_\mu(x)dx^\mu, \tag{3.31}$$

$$\omega^a{}_b(x) \equiv \omega^a{}_{b\mu}(x)dx^\mu, \tag{3.32}$$

that is, they are local 1-forms (see, e.g., Refs. [37]–[39]). This implies that all geometric properties of M can be expressed with these two 1-forms, their exterior products and their exterior derivatives only. Since exterior forms such as e^a and $\omega^a{}_b$ carry no coordinate indices (μ, ν, etc.), they are invariant under general coordinate transformations (diffeomorphisms) of M [40]. This means that a description of the geometry that only uses these forms is naturally coordinate-free. This is a welcome feature, since coordinates, like family names and street numbers are labels invented by humans, unrelated to geometry.

On the other hand, the 1-form exterior derivative operator, $d = dx^\mu \partial_\mu$, is such that acting on a p-form, α_p, yields the (p+1)-form, $d\alpha_p$. One of the fundamental properties of exterior calculus is that the second exterior derivative of a differential form vanishes identically,

$$d(d\alpha_p) =: d^2\alpha_p = dx^\mu \wedge dx^\nu \partial_\mu \partial_\nu \alpha \equiv 0. \tag{3.33}$$

This is trivially so because, when acting on continuously differentiable functions, the partial derivatives commute $\partial_\mu \partial_\nu \alpha = \partial_\nu \partial_\mu \alpha$, while the wedge product is antisymmetric, $dx^\mu \wedge dx^\nu = -dx^\nu \wedge dx^\mu$.

An additional — and very useful — simplification is brought about by the use of differential forms. Since all forms are invariant under diffeomorphisms, when applied to forms, the full covariant derivative $\mathfrak{D}$ is just D, and one can also forget about ∇. In other words, the Lorentz covariant derivative is the only derivative one has to deal with Therefore, only d and D will appear from now on, which is not only conceptually satisfactory, but also notationally welcome.

3.2.1   *Lorentz curvature*

A consequence of the identity $d^2\alpha = 0$ is that the square of the covariant derivative operator is not a differential operator, but an algebraic one. Consider the second covariant derivative of a vector field,

$$\begin{aligned}
D^2\phi^a &= D[d\phi^a + \omega^a{}_b \wedge \phi^b] \\
&= d[d\phi^a + \omega^a{}_b \wedge \phi^b] + \omega^a{}_b \wedge [d\phi^b + \omega^b{}_c \wedge \phi^c] \\
&= [d\omega^a{}_b + \omega^a{}_c \wedge \omega^c{}_b] \wedge \phi^b,
\end{aligned} \tag{3.34}$$

(here we have used the standard rule $d(\alpha \wedge \beta) = (d\alpha) \wedge \beta + (-1)^p \alpha \wedge d\beta$, where p is the degree of the form α [9]). The expression in brackets in the last equation is an antisymmetric second rank Lorentz tensor, the curvature two-form,

$$R^a{}_b = d\omega^a{}_b + \omega^a{}_c \wedge \omega^c{}_b \qquad (3.35)$$
$$= \frac{1}{2}R^a{}_{b\mu\nu}dx^\mu \wedge dx^\nu.$$

The connection $\omega^a{}_b(x)$ and the gauge potential in Yang-Mills theory, $\mathbf{A}(x) = \mathbf{A}_\mu dx^\mu$, are both 1-forms and have similar properties. This is no accident since they are both connections of a gauge group[3]; their transformation laws have the same form, and the curvature $R^a{}_b$ is completely analogous to the Yang-Mills field strength,

$$\mathbf{F} = d\mathbf{A} + \mathbf{A} \wedge \mathbf{A}. \qquad (3.36)$$

The fact that the two independent geometric ingredients, ω and e, play different roles is reflected by their distinct transformation rules under the Lorentz group: the vielbein transforms as a vector and not as a connection. In gauge theories vector fields play the role of matter, while connection fields represent interaction carriers.

3.2.2 Torsion

An important consequence of the different properties of the connection and the vielbein is that while a tensor two-form (the curvature) can be defined from ω and its exterior derivatives, it is impossible to construct a tensor two-form solely out of e^a and its derivatives. The only tensor obtained by differentiation of e^a is its covariant derivative, also known as the torsion 2-form,

$$De^a = de^a + \omega^a{}_b \wedge e^b = T^a, \qquad (3.37)$$

which involves both the vielbein and the connection. In contrast with T^a, the curvature $R^a{}_b$ is not a covariant derivative of some field.

Writing (3.37) explicitly in a coordinate basis gives

$$\partial_\mu e^a_\nu + \omega^a{}_{\mu b} \wedge e^b_\nu - (\mu \leftrightarrow \nu) = T^a_{\mu\nu}, \qquad (3.38)$$

where the torsion two-form is given by $T^a = (1/2)T^a_{\mu\nu}dx^\mu \wedge dx^\nu$. Since the symmetric part of this equation is arbitrary, it can also be written as

$$\partial_\mu e^a_\nu + \omega^a{}_{\mu b} \wedge e^b_\nu = \left(\frac{1}{2}\right)T^a_{\mu\nu} + \left(\frac{1}{2}\right)S^a_{\mu\nu}, \qquad (3.39)$$

where $S^a_{\mu\nu}$ is an arbitrary three-index *object*, symmetric in (μ, ν). The right hand side of this equation contains a hybrid object with indices $^a{}_{\mu\nu}$, which is *not necessarily a tensor* and that can be written as a partial projection of a three-index object in the base manifold,

$$\partial_\mu e^a_\nu + \omega^a{}_{\mu b} \wedge e^b_\nu = \left(\frac{1}{2}\right)[T^\lambda_{\mu\nu} + S^\lambda_{\mu\nu}]e^a_\lambda. \qquad (3.40)$$

---

[3]In the precise language of mathematicians, ω is "a locally defined Lie-algebra valued 1-form on M, which is also a connection in the principal $SO(D-1,1)$-bundle over M", while $\mathbf{A}$ is "a Lie-algebra valued 1-form on M, which is also a connection in the vector bundle of some gauge group G".

A moment's reflection allows to recognize this as the vielbein postulate (3.24), where the metric-compatible connection is

$$\Gamma^\lambda_{\mu\nu} = \left(\frac{1}{2}\right) T^\lambda_{\mu\nu} + \left(\frac{1}{2}\right) S^\lambda_{\mu\nu}. \tag{3.41}$$

The antisymmetric part given by the *torsion tensor* $T^\lambda_{\mu\nu}$, while the symmetric part $S^\lambda_{\mu\nu}$ is not determined by the definition of torsion. In order to fix S, an additional input is required and this is provided by the metricity postulate (3.29), which fixes it as,

$$\left(\frac{1}{2}\right) S^\lambda_{\mu\nu} = \frac{1}{2} g^{\lambda\rho} \left[\partial_\mu g_{\rho\nu} + \partial_\nu g_{\rho\mu} - \partial_\rho g_{\mu\nu}\right]. \tag{3.42}$$

3.2.3 *Riemann and Lorentz curvatures*

As we discussed in Section 3.1.3, the diffeomorphism group of the manifold and the Lorentz group of the tangent bundle define two independent connections and their corresponding covariant derivatives. The question then naturally arises as to the relationship between the Lorentz curvature two-form and the Riemann curvature tensor, $\widetilde{R}^\alpha{}_{\beta\mu\nu}$. In order to compare these two curvatures, the Riemann tensor can be projected on the tangent space to define the two-form

$$\widetilde{R}^a{}_b = \frac{1}{2} e^a{}_\alpha e^\beta{}_b \widetilde{R}^\alpha{}_{\beta\mu\nu} dx^\mu \wedge dx^\nu. \tag{3.43}$$

This expression has the same index structure as — and looks suspiciously similar to — the Lorentz curvature two-form (3.35). However, as we will see, these two objects coincide provided the geometry satisfies an additional condition, the vanishing of torsion. In order to see this, consider the definition of torsion (3.37). The equation for vanishing torsion $T^a = 0$ is an algebraic equation for ω. Let us call $\widetilde\omega$ the solution of this equation,

$$de^a + \widetilde\omega^a{}_b e^b = 0. \tag{3.44}$$

The torsion-free connection $\widetilde\omega(e, \partial e)$ is completely determined by metric structure of the manifold and can be expressed as

$$\widetilde\omega^a{}_{\mu b} = -E^\nu_b [\partial_\mu e^a_\nu - \Gamma^\lambda_{\mu\nu} e^a_\lambda], \tag{3.45}$$

where the connection $\Gamma^\lambda_{\mu\nu}$ is completely determined by the metric structure of the manifold, namely the Christoffel symbol (1.7). In fact, the attentive reader will recognize this expression as (3.40) for the case of vanishing torsion.

The difference between the Lorentz connection and the torsion-free connection is a tensor one-form known as the *contorsion tensor*,

$$\omega^a{}_b - \widetilde\omega^a{}_b = \kappa^a{}_b, \tag{3.46}$$

which is related to the torsion two-form, $T^a = \kappa^a{}_b e^b$. Then the Lorentz curvature two-form is

$$R^a{}_b = \widetilde{R}^a{}_b + \widetilde{D}\kappa^a{}_b + \kappa^a{}_c \kappa^c{}_b, \tag{3.47}$$

where $\widetilde{R}^a{}_b$ is the two-form defined by Riemann curvature in (3.43) and $\widetilde{D}$ is the covariant derivative for the torsion-free connection. Therefore, we conclude that the curvature two forms constructed with the Lorentz connection coincides with the Riemann curvature two-form if the torsion vanishes. Conversely, one can observe that the vanishing of R is quite independent from the vanishing of $\widetilde{R}$. In particular, it can be shown that a 2+1 Lorentz flat geometry, $R^a{}_b = 0$ implies a constant negative Riemann curvature, $\widetilde{R}^a{}_b = -\tau^2 e^a e_b$, where τ is an arbitrary constant with units of [length]$^{-1}$ [41].

3.2.4 *Building blocks*

As we saw previously, taking the second covariant derivative of a vector amounts to multiplying by the curvature 2-form. As a consequence, the covariant derivative of the curvature vanishes identically, an important property known as Bianchi identity,

$$DR^a{}_b = dR^a{}_b + \omega^a{}_c \wedge R^c{}_b - \omega^c{}_b \wedge R^a{}_c \equiv 0. \qquad (3.48)$$

This is an identity and not a set of equations; it is satisfied by *any* well defined connection 1-form whatsoever, and does not restrict in any way the form of the field $\omega^a{}_{b\mu}$ as it can be explicitly checked by substituting (3.35) in the second term of (3.48). If conditions $DR^a{}_b = 0$ were a set of equations instead, they would define a subset of connections that have a particular form, corresponding to some class of geometries.

As a consequence of this identity, taking successive covariant derivatives of e^a, $R^a{}_b$ and T^a, does not produce new independent tensors. For example, the covariant derivative of torsion gives

$$DT^a = dT^a + \omega^a{}_b \wedge T^b \equiv R^a{}_b \wedge e^b. \qquad (3.49)$$

Thus, the basic building blocks of a first order gravity action principle are:

e^a	Vielbein
$\omega^a{}_b$	Lorentz connection
$R^a{}_b$	Curvature two-form
T^a	Torsion
$\eta_{ab},\ \epsilon_{a_1 a_2 \cdots a_D}$	Invariant tensors

In order to describe a D-dimensional spacetime, the Lagrangian must be a D-form. Thus, with these objects and exterior products of them, a very limited number of locally Lorentz invariant actions can be put together in each dimension. For example, for $D = 4$ one can write

$$I[e, \omega] = \int \epsilon_{abcd}(\alpha R^{ab} e^c e^d + \beta e^a e^b e^c e^d), \qquad (3.50)$$

which is nothing but the Einstein-Hilbert action supplemented by a cosmological constant term (3.1) written in terms of differential forms. Other expressions as for example the Gauss-Bonnet term,

$$I[e, \omega] = \int \epsilon_{abcde} R^{ab} R^{cd} e^e, \tag{3.51}$$

defines a suitable Lagrangian in five dimensions. In these two cases the local Lorentz symmetry is manifest by the fact that the integrands are Lorentz invariant. Under an infinitesimal local Lorentz transformation with parameter λ^a_b the dynamical fields transform as

$$\delta_\lambda e^a = \lambda^a{}_b e^b, \qquad \delta_\lambda \omega^a{}_b = -D\lambda^a{}_b. \tag{3.52}$$

From definition (3.35), it can be seen that the change in the curvature produced by a first order change in ω is $\delta R^a{}_b = D\delta\omega^a{}_b$. It is then a simple exercise to show explicitly that under these transformations the actions (3.50) and (3.51) are invariant. As previously mentioned, the Lorentz transformations of the connection can be written as a gauge transformation $\delta_\lambda A = d\lambda + [A, \lambda]$, with

$$A = \frac{1}{2}\omega^{ab} J_{ab}, \tag{3.53}$$

where J_{ab} represent the Lorentz generators and $\lambda = \frac{1}{2}\lambda^{ab} J_{ab}$ parametrizes an infinitesimal Lorentz transformation. Later, we will see that extending the Lorentz group to a larger group allows to include the vielbein in the same footing as the Lorentz connection as part of an extended connection.

3.3 Gravity as a Gauge Theory

Symmetry principles help in constructing the right classical action and, more importantly, they are often sufficient to ensure the viability of the quantum theory obtained from its classical limit. In particular, as discussed at the very beginning of this book, gauge symmetry is the key to prove consistency (renormalizability) of the quantum field theory that provides the best description so far for three of the four basic interactions of nature. The gravitational interaction — at least in its simplest version — has stubbornly escaped this pattern in spite of the fact that, as we saw, it is described by a theory endowed with general covariance, which is a local invariance quite analogous to gauge symmetry. Here we try to shed some light on this puzzle.

Roughly one year after Yang and Mills proposed their model for non-abelian gauge invariant interactions [7], R. Utiyama showed that the Einstein theory can also be written as a gauge theory for the Lorentz group [18]. As shown in the previous section, this can be checked directly from the Lagrangian in (3.50), which is a Lorentz scalar and hence, trivially invariant under (local) Lorentz transformations. This generated the expectation to construct gravity as a gauge theory for the Poincaré group, $G = ISO(3,1)$, which is the standard symmetry group in particle physics, that includes both Lorentz transformations and translations. The inclusion of translations seems natural in view of the fact that a general coordinate transformation

$$x^\mu \to x'^\mu = x^\mu + \xi^\mu(x), \qquad (3.54)$$

looks like a local translation. This is undoubtedly a local symmetry in the sense that it is a transformation that leaves the action invariant and whose parameters ξ are functions of the position in spacetime. This suggests that diffeomorphism invariance could be identified with a symmetry under local translations that extends the Lorentz group into its Poincaré embedding. Although this looks plausible, a local action for a Poincaré connection, invariant under (3.54), has not been found so far. The problem is how to implement this symmetry as a gauge transformation of the dynamical fields (e, ω or $g_{\mu\nu}$). Under a coordinate diffeomorphism (3.54), the metric changes by a Lie derivative,

$$\begin{aligned} \delta g_{\mu\nu}(x) &= g_{\mu\nu}(x + \xi(x)) - g_{\mu\nu}(x) \\ &= -g_{\alpha\nu}\left[\partial_\mu \xi^\alpha + \Gamma^\alpha_{\mu\lambda}\xi^\lambda\right] - (\mu \leftrightarrow \nu), \end{aligned} \qquad (3.55)$$

where Γ is the Christoffel symbol, which is the appropriate connection for general coordinate diffeomorphisms. However, the Christoffel symbol is not a fundamental field in the so-called *second order formalism* in which the metric is the fundamental field. In the *first order formalism* on the other hand, in which the fundamental fields are the vielbein and the Lorentz connection, the action is trivially invariant under (3.54), precisely because the differential forms e and ω are coordinate invariant and there is no other naturally defined object that plays the role of a connection for diffeomorphisms.

Attempts to identify the coordinate transformations with local translations, have also systematically failed. The problem is that there is no local four-dimensional action for a connection $\mathbf{A}$, invariant under local $ISO(3,1)$ transformations [42–45]. Although the fields ω^{ab} and e^a have the right tensor properties to match the generators of the Poincaré group, there is no Poincaré-invariant 4-form available constructed with the connection for the Lie algebra of $ISO(3,1)$. In spite of appearances, the superficially correct assertion that gravity is a gauge theory for the translation group is crippled by the profound differences between a gauge theory with fiber bundle structure and another with an open algebra, such as gravity.

The best way to convince ourselves of this obstruction is by assuming the existence of a connection for the group of local translations in the tangent space analogous to the spin connection. The new gauge field should have an index structure to match that of the generator of translations, P_a in the Poincaré algebra,

$$\begin{aligned} [P_a, P_b] &= 0, \qquad [J_{ab}, P_c] = \eta_{bc}P_a - \eta_{ac}P_b, \\ [J_{ab}, J_{cd}] &= \eta_{bc}J_{ad} - \eta_{ac}J_{bd} + \eta_{ad}J_{bc} - \eta_{bd}J_{ac}. \end{aligned} \qquad (3.56)$$

The gauge theory for the Poincaré group should be based on a connection that generalizes the Lorentz connection (3.53),

$$\mathbf{A} = s^a P_a + \frac{1}{2}\omega^{ab} J_{ab} \,. \tag{3.57}$$

where the new connection field s would be a one-form and a vector under Lorentz transformations. The challenge is to construct an invariant action for the fields ω^{ab} and s^a, which transform under translations as

$$\delta\omega^{ab} = 0\,, \qquad \delta s^a = d\lambda^a + \omega^a{}_b \lambda^b \equiv D\lambda^a. \tag{3.58}$$

The connection piece s^a has the same transformation properties under the Lorentz group — and same index structure — as the vielbein. If no other fields are included, the four-dimensional Lagrangian $L[s,\omega]$ should be a four form constructed out of s^a, $\omega^a{}_b$, their first exterior derivatives and the invariant tensors η_{ab} and ϵ_{abcd}. A quick check shows that the only four-form Lorentz-invariant candidates are

$$\epsilon_{abcd}s^a s^b s^c s^d,\ \epsilon_{abcd}R^{ab}s^c s^d,\ \epsilon_{abcd}R^{ab}R^{cd},\ \eta_{ac}\eta_{bd}R^{ab}s^c s^d,\ \eta_{ab}S^a S^b, \tag{3.59}$$

where $S^a := ds^a + \omega^a{}_b s^b = Ds^a$. It can be easily checked that while these four-forms are identically Lorentz invariant, under (3.58) they are not, unless $S^a = 0$. There is one exceptional combination that is invariant,

$$\sigma = \eta_{ac}\eta_{bd}R^{ab}s^c s^d + \eta_{ab}S^a S^b, \tag{3.60}$$

which is trivially invariant because it is a total derivative (exact form),

$$\sigma = d(\eta_{ab}S^a s^b). \tag{3.61}$$

The reader may suspect by now why the Einstein action (3.50) — with or without the cosmological constant β, is not invariant under local Poincaré translations, where $\delta_\lambda e^a = D\lambda^a$ and $\delta_\lambda \omega^{ab} = 0$. The point is that, modulo an exact form,

$$\delta_\lambda[\epsilon_{abcd}R^{ab}e^c e^d] = 2\epsilon_{abcd}R^{ab}\lambda^c T^d, \tag{3.62}$$

which does not vanish for arbitrary Riemann curvature unless one assumes $T^a = 0$. Hence, the local Poincaré translation is at best an on-shell symmetry of the standard Einstein-Hilbert action.[4]

A more sophisticated approach could be to replace the Poincaré group by allowing any group G that contains the Lorentz transformations as a subgroup. The idea is as follows: Our conviction that the space we live in is four-dimensional and approximately flat results from our experience that we can act with the group of four-dimensional translations to connect any two points in spacetime. But we know this to be only approximately true. Like the surface of the Earth, our spacetime

---

[4]Some attempts to improve this ansatz introduce an auxiliary non-dynamical field, similar to the Stueckelberg compensating field in gauge theory, that renders the vielbein translation-invariant [46]. At the end of the day, however, the resulting gauge theory is one where the translation symmetry is "spontaneously broken", which points that this purported symmetry was never there.

could be curved but with a radius of curvature so large we wouldn't notice the deviation from flatness except in very delicate observations. For instance, instead of the symmetries of a four-dimensional *flat* spacetime, we might be experiencing the symmetries of a four-dimensional spacetime of *nonzero constant curvature*, also known as a pseudosphere.

The smallest nontrivial choices for G, which are not just a direct product of the form $SO(3,1) \times G_0$, are:

$$G = \begin{cases} SO(4,1) & \text{de Sitter (dS)} \\ SO(3,2) & \text{anti-de Sitter (AdS)} \\ ISO(3,1) & \text{Poincaré.} \end{cases} \tag{3.63}$$

Both de Sitter and anti-de Sitter groups are semi-simple, while the Poincaré group is not — it can be obtained as a contraction of either dS or AdS. This technical detail could mean that $SO(4,1)$ and $SO(3,2)$ have better chances than the Poincaré group to become physically relevant for gravity. Semi-simple groups are preferred as gauge groups because they have an invariant in the group, known as the *Killing metric*, which can be used to define kinetic terms for the gauge fields.[5]

In spite of this improved scenario, it is still impossible to express gravity in four dimensions as a gauge theory for the dS or AdS groups. However, as we shall see next, in odd dimensions ($D = 2n - 1$), and only in that case, gravity can be cast as a gauge theory of the groups $SO(D,1)$, $SO(D-1,2)$, or $ISO(D-1,1)$, in contrast with what one finds in dimension four, or in any other even dimension.

---

[5]Non semi-simple groups contain *abelian* invariant subgroups. The generators of the abelian subgroups commute among themselves, and the fact that they are invariant subgroups implies that too many structure constants in the Lie algebra vanish. This in turn makes the Killing metric to acquire zero eigenvalues preventing its invertibility.

Chapter 4

Gravity in Higher Dimensions

We now turn to the construction of an action for D-dimensional gravity as a local functional of the one-forms e^a, ω^a_b and their exterior derivatives. The fact that $d^2 \equiv 0$ implies that the Lagrangian must involve at most first derivatives of these fields through the two-forms R^a_b, T^a, together with the two invariant tensors of the Lorentz group, η_{ab}, and $\epsilon_{a_1 \cdots a_D}$ used to raise, lower and contract indices. We need not worry about invariance under general coordinate transformations as exterior forms are coordinate scalars by construction. On the other hand, the action principle cannot depend on the choice of basis in the tangent space and hence Lorentz invariance should be respected. A sufficient condition to ensure Lorentz invariance is to demand the Lagrangian to be a Lorentz scalar although, this is sufficient and may not be strictly necessary.

Finally, since the action must be an integral over the D-dimensional space-time manifold, the problem is to construct a D-form with the following ingredients:

$$e^a, \quad \omega^a_b, \quad R^a_b, \quad T^a, \quad \eta_{ab}, \quad \epsilon_{a_1 \cdots a_D}. \tag{4.1}$$

Hence, we tentatively postulate the Lagrangian for gravity to be a D-form made of linear combinations of manifestly Lorentz invariant products of these ingredients. We do not include other fields such as the metric, that are not elementary objects, nor exterior products or exterior derivatives thereof to deserve their inclusion in the action as independent fields. This rule excludes the inverse metric and the Hodge $\star$-dual. This rule is additionally justified since it explicitly excludes inverse fields, like $E^\mu_a(x)$, which would be like A^{-1}_μ in Yang-Mills theory (see Refs. [47] and [48] for more on this). This postulate also rules out the inclusion of tensors like the Ricci tensor $R_{\mu\nu} = e^\lambda_a \eta_{bc} e^c_\mu R^{ab}_{\lambda\nu}$, or $R_{\alpha\beta} R_{\mu\nu} R^{\alpha\mu\beta\nu}$, etc. Moreover, the resulting theories of gravity in arbitrary D dimensions are sufficiently sensible, and reduce to the standard GR case that passes all the observational tests in $D = 4$.

4.1   Lovelock Lagrangians

The natural extension of the Einstein-Hilbert action for $D > 4$ which is also valid for $D \leq 4$ is provided by the following.

Theorem [Lovelock, 1970 [49]-Zumino, 1986 [47]]: The most general Lorentz-invariant action for gravity that does not involve torsion and gives at most second order field equations for the metric under the standard assumption of vanishing torsion,[1] has the form

$$I_D = \int_M \sum_{p=0}^{[D/2]} a_p L^{(D,p)}, \tag{4.2}$$

where the a_p's are arbitrary constants, and $L^{(D,p)}$ is given by[2]

$$L^{(D,\,p)} = \epsilon_{a_1 \cdots a_D} R^{a_1 a_2} \cdots R^{a_{2p-1} a_{2p}} e^{a_{2p+1}} \cdots e^{a_D}. \tag{4.3}$$

Proof: Clearly, the Lorentz invariants $L^{(D,\,p)}$ are allowed combinations of the ingredients in (4.1), and a moment's reflection shows that Lorentz invariants involving $\epsilon_{a_1 \cdots a_D}$ can only contain products of Rs and es. Since these ingredients have as many Lorentz indices as their form degree, these products could only involve p curvatures and $D - 2p$ vielbeins that saturate all the indices of the Levi-Civita symbol, precisely as in the form $L^{(D,\,p)}$. As will be shown in Section 4.2, direct variation yields field equations that are again combinations of e^a, $R^a{}_b$ and T^a. As seen in Section 3.2.3, in the assumption of vanishing torsion, the connection can be expressed in terms of the metric, and the curvature two-form is given by the Riemann curvature $\widetilde{R}^{ab}$, which involves up to second derivatives of the metric. Thus, the field equations obtained from the Lovelock action (4.2) involve up to second derivatives of the metric.

But, are (4.3) the only invariants that one should consider? If torsion is not involved T^a must clearly be removed from the list of ingredients (4.1). Additionally, in order to produce Lorentz invariant expressions the connection should only occur in the combination $R^a{}_b = d\omega^a{}_b + \omega^a{}_c \omega^c{}_b$. Thus, the remaining ingredients are e^a, $R^a{}_b$ and η_{ab}. Next, we can observe that contractions using η_{ab} like $\eta_{ab} e^a e^b$ and $\eta_{ab} e^a R^{bc}$ must also be ruled out if torsion is not involved (the first vanishes identically and the second equals $-DT^c$). Therefore, η_{ab} can only be used to contract curvature two-form with itself in expressions like $P_k = R^{a_1}{}_{a_2} R^{a_2}{}_{a_3} \cdots R^{a_k}{}_{a_1}$. These Lorentz scalar $2k$-forms can be multiplied with other of the same type to produce a possible Lagrangian D-form, as for instance,

$$P_{k_1} P_{k_2} \cdots P_{k_s}, \quad \text{with } k_1 + k_2 + \cdots + k_s = D/2. \tag{4.4}$$

However, as will be discussed in Section 4.3, these expressions, that only exist in dimensions $D = 4, 8, \ldots, 4n$, are topological invariant densities –locally exact

---

[1]These conditions can be translated to mean that the Lovelock theories possess the same degrees of freedom as the Einstein Hilbert action in each dimension, that is, $D(D-3)/2$ (see, e.g., Ref. [50])
[2]From now on we omit the wedge symbol $\wedge$ in the exterior products.

forms– and do not give rise to local field equations. Thus, we conclude that in the assumption of vanishing torsion the only invariants that can be included in the Lagrangian are the D-forms $L^{(D,\,p)}$ and linear combinations thereof **(Q.E.D.)**

Let us examine now what this theorem means in different dimensions.

- **D = 2**: In two dimensions the action reduces to a linear combination of the 2-dimensional Euler character, χ_2, and the spacetime volume (area),

$$I_2[e,\omega] = \int_M \epsilon_{ab}[a_1 R^{ab} + a_0 e^a e^b]$$

$$= \int_M \sqrt{|g|}\,(a_1 R + 2a_0)\,d^2x \qquad (4.5)$$

$$= a_1 \cdot \chi_2 + 2a_0 \cdot V_2.$$

Varying this action yields

$$\delta I_2 = \int_M \left[a_1 d(\epsilon_{ab}\delta\omega^{ab}) + 2a_0\epsilon_{ab}\delta e^a e^b\right]. \qquad (4.6)$$

The first term on the right can be written as an integral over boundary of M, and leads to no equations for ω in M. This reflects the fact that the Euler characteristic χ_2, being a topological invariant, is an integer that does not change under continuous deformations as those explored by the variation. In other words, χ_2 is stationary under arbitrary continuous deformations of ω. On the other hand, demanding that the action be stationary under arbitrary independent variations of e^a implies that the vielbein vanish everywhere in M. This is certainly an extremum of the volume form, although not a very interesting one.

Thus, unless some other fields like matter sources are included, I_2 does not make a very useful dynamical theory for geometry. If the manifold M has Euclidean metric and a fixed boundary, a boundary term must be added and the extreme of the action is a minimal surface in the shape of a soap bubble. This is the famous *Plateau's problem*, which consists of establishing the shape of the surface of minimal area bounded by a given closed curve [51].

- **D = 3**: In this case the expression (4.2) reduces to the Hilbert action with a volume term, whose coefficient is the cosmological constant. This is the standard generalization of GR for $D = 3$,

$$I_3[e,\omega] = \int_M \epsilon_{abc}[a_1 R^{ab}e^c + a_0 e^a e^b e^c]$$

$$= -\int_M \sqrt{|g|}\,(a_1 R + 6a_0)\,d^3x. \qquad (4.7)$$

The field equations are

$$\delta\omega \rightarrow \epsilon_{abc}T^c = 0\,, \quad \delta e \rightarrow \epsilon_{abc}[a_1 R^{ab} + 3a_0 e^a e^b] = 0\,, \qquad (4.8)$$

which describe a torsion-free geometry with constant negative curvature, *three-dimensional anti-de Sitter space*, or AdS_3. Note that, although we have not included torsion in the action, we have not demanded it to vanish identically, as would be the case in the purely metric formulation of gravity. Here the affine structure (connection) and the metric structure (vielbein) are considered dynamically independent and therefore independently varied. The vanishing of torsion is a dynamical equation, an output of the action principle. This approach, sometimes referred to as the *Palatini formulation*, contains the same solutions and is completely equivalent to the metric one provided there are no other dynamical fields that couple to ω. We will come back to this in Section 4.2.2.

• **D = 4**: The Lovelock action contains, in addition to the Einstein-Hilbert and cosmological terms, the four-dimensional Euler form χ_4,

$$
\begin{aligned}
I_4 &= \int_M \epsilon_{abcd} \left[a_2 R^{ab} R^{cd} + a_1 R^{ab} e^c e^d + a_0 e^a e^b e^c e^d \right] \\
&= -\int_M \sqrt{|g|} \left[a_2 \left(R^{\alpha\beta\gamma\delta} R_{\alpha\beta\gamma\delta} - 4 R^{\alpha\beta} R_{\alpha\beta} + R^2 \right) + 2 a_1 R + 24 a_0 \right] d^4 x \\
&= -a_2 \cdot \chi_4 - 2 a_1 \int_M \sqrt{|g|} R \, d^4 x - 24 a_0 \cdot V_4 .
\end{aligned}
\tag{4.9}
$$

Again the variation with respect to ω and e yield

$$
\delta\omega \to \epsilon_{abcd} e^c T^d = 0 , \quad \delta e \to \epsilon_{abcd} [a_1 R^{ab} + 2 a_0 e^a e^b] e^b = 0 .
\tag{4.10}
$$

The first of these is again the statement that the geometry is torsion-free; the second is just Einstein's equations, in exterior form. Since on shell the torsion vanishes, one can set $\tilde{R}^{ab} = R^{ab}$.

• **D > 4**: For all dimensions, the Lagrangian is a polynomial of degree $p \leq D/2$ in the curvature 2-form. In general, each term $L^{(D,\, p)}$ in the Lagrangian (4.2) is the dimensional continuation to D dimensions of the Euler density from all even dimension below D [47]. In particular, for $D = 2n$ the highest power in the curvature is the *Euler $2n$-form*,

$$
\mathfrak{E}_{2n} = \epsilon_{a_1 b_1 a_2 b_2 \cdots a_n b_n} R^{a_1 b_1} R^{a_2 b_2} \cdots R^{a_n b_n} .
\tag{4.11}
$$

In four dimensions, the term $L^{(4,\, 2)} = \mathfrak{E}_4$ in (4.9) is also known as the *Gauss-Bonnet density*, whose integral over a closed compact four dimensional manifold M_4 equals the *Euler characteristic* $\chi(M_4)$. By dimensionally continuing to five dimensions provides the first nontrivial generalization of Einstein gravity allowed by Lovelock's theorem, the *Gauss-Bonnet 5-form*

$$
\epsilon_{abcde} R^{ab} R^{cd} e^e = -\sqrt{|g|} \left[R^{\alpha\beta\gamma\delta} R_{\alpha\beta\gamma\delta} - 4 R^{\alpha\beta} R_{\alpha\beta} + R^2 \right] d^5 x .
\tag{4.12}
$$

It has been known for many years that this term could be added to the Einstein-Hilbert action in five dimensions. This is commonly attributed to Lanczos [52], but the original source is unclear.

The Lovelock Lagrangian was also identified as describing the only ghost-free[3] effective theory for a spin-two field, obtained from string theory in the low energy limit [47, 53]. From our perspective, the absence of ghosts is only a reflection that the Lovelock action yields at most second order field equations for the metric, so that the propagators behave as k^{-2}, and not as $k^{-2} + \alpha k^{-4}$, as would be the case in a general higher derivative theory.

4.2   Dynamical Content

4.2.1   *Field equations*

Varying the action (4.2) with respect to e^a and ω^{ab} yields, modulo surface terms,

$$\delta I_D = \int [\delta e^a \mathcal{E}_a + \delta \omega^{ab} \mathcal{E}_{ab}] = 0, \tag{4.13}$$

where

$$\mathcal{E}_a = \sum_{p=0}^{[\frac{D-1}{2}]} a_p (D - 2p) \mathcal{E}_a^{(p)}, \tag{4.14}$$

and

$$\mathcal{E}_{ab} = \sum_{p=1}^{[\frac{D-1}{2}]} a_p p (D - 2p) \mathcal{E}_{ab}^{(p)}, \tag{4.15}$$

and we have defined

$$\mathcal{E}_a^{(p)} := \epsilon_{ab_2 \cdots b_D} R^{b_2 b_3} \cdots R^{b_{2p} b_{2p+1}} e^{b_{2p+2}} \cdots e^{b_D}, \tag{4.16}$$

$$\mathcal{E}_{ab}^{(p)} := \epsilon_{aba_3 \cdots a_D} R^{a_3 a_4} \cdots R^{a_{2p-1} a_{2p}} T^{a_{2p+1}} e^{a_{2p+2}} \cdots e^{a_D}. \tag{4.17}$$

The condition for I_D to have an extremum under arbitrary first order variations is that $\mathcal{E}_a$ and $\mathcal{E}_{ab}$ vanish. These equations involve only first derivatives of e^a and $\omega^a{}_b$, simply because $d^2 = 0$. If one further assumes, as usual, that the torsion identically vanishes, then (4.15) is automatically zero. In addition, as mentioned in the previous chapter, the torsion-free condition $T^a = 0$ can be solved for ω as a function of the vielbein, its inverse (E_a^μ) and its derivative as in (3.45)

$$\widetilde{\omega}^a{}_{b\mu} = -E_b^\nu (\partial_\mu e_\nu^a - \Gamma_{\mu\nu}^\lambda e_\lambda^a), \tag{4.18}$$

---

[3]Physical states in quantum field theory have positive probability, which means that they are described by genuine vectors in a Hilbert space. Ghosts instead, are unphysical negative norm states. A Lagrangian containing arbitrarily high derivatives of fields generally leads to ghosts. It is unexpected and highly nontrivial that a gravitational action such as (4.2) leads to a ghost-free theory.

where $\Gamma^\lambda_{\mu\nu}$ is the torsion-free affine connection, symmetric in $\mu\nu$ and given by the Christoffel symbol (3.42). Substituting expression (4.18) for the spin connection back into (4.16) yields second order field equations for the metric. These equations are identical to those obtained from the Lovelock action written in terms of the Riemann tensor and the metric,

$$I_D[g] = \int_M d^D x \sqrt{|g|} \left[a_0' + a_1' \widetilde{R} \right.$$
$$\left. + a_2' \left(\widetilde{R}^{\alpha\beta\gamma\delta} \widetilde{R}_{\alpha\beta\gamma\delta} - 4\widetilde{R}^{\alpha\beta} \widetilde{R}_{\alpha\beta} + \widetilde{R}^2 \right) + \cdots \right]. \qquad (4.19)$$

This purely metric construction, also called *second order formalism* yields actions that involve the metric and its first and second derivatives. A Lagrangian containing ∂g and $\partial^2 g$ can yield up to fourth order equations for the metric, as in the case of $f(R)$ theories or for Lagrangians involving arbitrary contractions of the Riemann tensor. It is the magic of the Lovelock actions in general- and of the Einstein-Hilbert action in particular, that the field equations turn out to be second order only. One way to understand this is by observing that all the second derivatives in the Lagrangian can be collected in a boundary term (total derivative), which can be safely neglected. This boundary term can be identified by direct computation in the Einstein-Hilbert case, but it can become a formidable task for $L(D, p)$ with arbitrarily large p.

Some authors refer to the fact that the Lovelock actions contain higher powers of the curvature as *higher derivative theories of gravity*, which is clearly incorrect. Higher derivatives equations for the metric would mean that the initial conditions required to uniquely determine the time evolution are not those of General Relativity and hence the theory would have more degrees of freedom than standard gravity [54]. Higher order equations would also make the propagators in the quantum theory to develop poles at imaginary energies: *ghosts*. Ghost states spoil the unitarity of the theory, making it hard to interpret its meaning and to trust its predictions.

4.2.2 *First and second order theories*

Lovelock theories for $D \leq 4$ and for $D > 4$ behave very differently. In the first case the field equations (4.14) and (4.15) are linear in R^{ab} and T^a, while in the latter case the equations are nonlinear in the curvature tensor. In particular, while for $D \leq 4$ the equations (4.17) imply the vanishing of torsion, this is no longer true for $D > 4$. In fact, the field equations evaluated in certain configurations may leave some components of the curvature and torsion undetermined. For example, Eq. (4.15) has the form of a polynomial $P(R^{ab})$ times T^a, and it is possible that for certain forms of the geometry P vanish, leaving the torsion tensor completely undetermined. The configurations for which the equations do not determine R^{ab} and T^a form a set of measure zero in the space of geometries, while in generic cases, outside of these degenerate configurations, the Lovelock theory has the same

$D(D-3)/2$ degrees of freedom as ordinary gravity [55]. Although this can be seen as a sign that one should not worry too much about potential degeneracies, the problem is rather serious.

Assuming torsion to be identically zero means that e^a and $\omega^a{}_b$ are not independent fields, contradicting the assumption that these fields correspond to two independent features of the geometry. However, for $D \leq 4$, (4.17) is proportional to T^a, therefore imposing the torsion-free constraint is at best unnecessary. On the other hand, for $D > 4$, the equation $\mathcal{E}_{ab} = 0$ does not necessarily imply $T^a = 0$. Thus the torsion free-condition is a more drastic condition on the dynamics than the field equations; it is a truncation of the theory.

The equivalence between the first order formulation and the metric one rests on the fact that the equation obtained varying the action with respect to ω implies $T^a = 0$, which is an algebraic equation for the connection that can in principle be solved for the same field, the solution being $\tilde{\omega}$. In general, if the field equation for some field ϕ can be solved algebraically as $\phi = f(\psi)$ in terms of the remaining fields, then by the implicit function theorem, the original action principle $I[\phi, \psi]$ is identical to the reduced one obtained by substituting $f(\psi)$ in the action, $I[f(\psi), \psi]$. This occurs in 3 and 4 dimensions, where the spin connection can be algebraically obtained from its own field equation and $I[\omega, e] = I[\omega(e, \partial e), e]$. In higher dimensions, however, the torsion-free condition is not necessarily a consequence of the field equations. Although (4.17) is algebraic in ω, it is practically impossible to solve for ω as a function of e. Therefore, it is not clear in general whether the action $I[\omega, e]$ is equivalent to the second order form of the action, $I[\omega(e, \partial e), e]$.

Matters get worse if other fields couple to the connection, as it occurs for instance when fermions are present. Varying the action with respect to ω produces extra contributions, the torsion no longer vanishes and therefore ω cannot be expressed purely in terms of the metric. A more serious problem arises if other fields couple to the curvature, as in the conformal coupling of a scalar field, $\phi^k R$, because then the variation with respect to the connection no longer yields an algebraic relation that can be solved for ω. In all of these systems, torsion does not vanish in the first order formulation while the second order formulation assumes $T^a \equiv 0$. These are not just two different formalisms of the same system; they are two different dynamical theories.[4]

So far, this discussion has been limited to the classical theory. If one wishes to extend the metric formulation to the quantum regime, it seems highly improbable that it could work even in $D = 4$, where the connection is not the fundamental field, but a shorthand expression for a complicated function of the metric, which is assumed to be the true fundamental field.

---

[4]It would be more appropriate to call these two schemes *first and second order theories* because they are not two equivalent or complementary viewpoints to discuss the same dynamical system.

4.3   Torsional Series

As already mentioned, the torsion-free condition does not automatically follow from the field equation (4.15). It could be that the torsion is completely indeterminate, as it happens for instance if the geometry has constant curvature of a certain radius that depends on the choice of a_p's to be discussed below. Thus, even if torsion is not included in the Lagrangian it might appear in the classical equations.

It seems therefore natural to consider the generalization of the Lovelock Lagrangians in which torsion *is not* assumed to vanish identically. This generalization includes all possible Lorentz invariants involving those terms that were excluded on the assumption of $T^a = 0$. This includes combinations like $R^{ab}e_b$, which do not involve torsion explicitly but were excluded because $R^{ab}e_b = DT^a$.

The general construction was worked out in [56]. The main difference with the torsion-free case is that now the invariant Levi-Civita tensor $\epsilon_{a_1 a_2 \cdots a_D}$ is not used and the new terms are contractions that only involve η_{ab}. This means that, apart from the dimensional continuation of the Euler densities, expressions related to the Pontryagin (or Chern) classes are also included.

For $D = 3$, the only new torsion term not included in the Lovelock family is

$$e^a T_a, \tag{4.20}$$

while for $D = 4$, there are three terms not included in the Lovelock series,

$$e^a e^b R_{ab}, \quad T^a T_a, \quad R^{ab} R_{ab}. \tag{4.21}$$

The last term in (4.21) is the Pontryagin density, whose integral also yields a topological invariant. A linear combination of the other two terms is a topological invariant known as the *Nieh-Yan form*, given by [57]

$$N_4 = T^a T_a - e^a e^b R_{ab}. \tag{4.22}$$

The properly normalized integral of (4.22) over a 4-manifold is an integer related to the $SO(5)$ and $SO(4)$ Pontryagin classes [58].

In general, the torsional terms that can be included in the action are invariants that can take any of these forms:

$$A_{2n} = e_{a_1} R^{a_1}_{a_2} R^{a_2}_{a_3} \cdots R^{a_{n-1}}_{a_n} e^{a_n}, \quad \text{even } n \geq 2, \tag{4.23}$$

$$B_{2n+1} = T_{a_1} R^{a_1}_{a_2} R^{a_2}_{a_3} \cdots R^{a_{n-1}}_{a_n} e^{a_n}, \quad \text{any } n \geq 1, \tag{4.24}$$

$$C_{2n+2} = T_{a_1} R^{a_1}_{a_2} R^{a_2}_{a_3} \cdots R^{a_{n-1}}_{a_n} T^{a_n}, \quad \text{odd } n \geq 1. \tag{4.25}$$

These are Lorentz invariant $2n$, $2n+1$ and $2n+2$ forms, respectively. They belong to the same family as the Pontryagin densities or Chern classes,

$$P_{2n} = R^{a_1}_{a_2} R^{a_2}_{a_3} \cdots R^{a_n}_{a_1}, \quad \text{even } n. \tag{4.26}$$

The trace of an odd number of curvature two-forms vanishes identically due to the antisymmetry $R^{ab} = -R^{ba}$. Similar symmetry arguments yield the restrictions for n appearing in (4.23)–(4.25).

The Lagrangians that can be constructed now have a greater diversity and there is no uniform expression that can be provided for all dimensions. For example in 8 dimensions, in addition to the Lovelock terms, all possible 8-forms made by taking products among the elements of the set $\{A_4, A_8, B_3, B_5, B_7, C_4, C_8, P_4, P_8\}$ can be included. There are ten such combinations:

$$(A_4)^2, A_8, B_3 B_5, A_4 C_4, (C_4)^2, C_8, A_4 P_4, C_4 P_4, (P_4)^2, P_8. \tag{4.27}$$

To make life even harder, there are some linear combinations of these products like (4.22) which are topological densities. In 8 dimensions there are two Pontryagin forms

$$P_8 = R^{a_1}_{\ a_2} R^{a_2}_{\ a_3} R^{a_3}_{\ a_4} R^{a_4}_{\ a_1},$$
$$(P_4)^2 = (R^a_{\ b} R^b_{\ a})^2,$$

which also occur in the absence of torsion, and there are two generalizations of the Nieh-Yan form,

$$(N_4)^2 = (T^a T_a - e^a e^b R_{ab})^2,$$
$$N_4 P_4 = (T^a T_a - e^a e^b R_{ab})(R^c_{\ d} R^d_{\ c}).$$

For different dimensions the number $\mathcal{N}(D)$ of nonvanishing torsional terms varies enormously, as shown in the following table [56].

Table 4.1

D	3	4	5	6	7	8	9	10	11	12	13	14	15	16	17	18	19	20	21	22	23	24	25	26
$\mathcal{N}$(D)	1	3	1	0	4	10	4	1	13	27	13	5	36	69	36	18	91	161	92	53	213	361	140	217

Looking at these expressions one can easily feel depressed. The Lagrangians look awkward and the number of terms in them grows wildly with the dimension. Indeed, as it is shown in [56], $\mathcal{N}(D)$ grows as $p(D/4)$, where $p(N)$ is the partition function of N, defined as the number of ways the integer N can be written as the sum of positive integers. For large D this number is given by the *Hardy-Ramanujan formula*, $p(D/4) \sim \frac{1}{\sqrt{3D}} \exp[\pi \sqrt{D/6}]$ [59].

Still, this number is small compared with what is allowed if the restriction that these invariants be written as exterior products of the basic ingredients T^a, R^{ab} and e^a is lifted. In fact, the number of invariant contractions of the torsion tensor, the Riemann tensor and the metric is infinite; to be sure, the number of functions $f(R)$ that can be considered in the metric approach is already infinite.

The proliferation of terms in the action in higher dimensions is not a purely aesthetic problem. The coefficients in front of each term in the Lagrangian are arbitrary and dimensionful. This problem already occurs in 4 dimensions, where Newton's constant and the cosmological constant have dimensions of [length]^2 and [length]$^{-4}$ respectively and their relative weight is not theoretically fixed. As evidenced by the outstanding cosmological constant problem, there is no theoretical prediction for its value in a way that can be compared with observations.

4.4   Born-Infeld Lagrangians

There are no gravitation theories with enhanced symmetry from $SO(D-1,1)$ to $SO(D,1)$, $SO(D-1,2)$ or $ISO(D-1,1)$ for $D = 2n$ dimensions. Any choice of Lovelock coefficients seems as good as any other, but the possibility of having a unique maximally symmetric vacuum is obtained in three cases only, corresponding to dS, AdS and Minkowski. The first two cases in even dimensions are obtained with the so-called Born-Infeld (**BI**) theories [60–62], while the third is produced by the Einstein-Hilbert Lagrangian. The BI Lagrangian is given by a particular choice of the a_p's so that the Lovelock Lagrangian takes the form

$$L_{2n}^{BI} = \epsilon_{a_1 \cdots a_{2n}} \bar{R}^{a_1 a_2} \cdots \bar{R}^{a_{2n-1} a_{2n}}, \tag{4.28}$$

where $\bar{R}$ also known as the *concircular curvature* [63], is given by

$$\bar{R}^{ab} := R^{ab} \pm \frac{1}{l^2} e^a e^b, \tag{4.29}$$

The expression (4.28) is the Pfaffian form for the concircular curvature which can be seen as the gravitational analogue of the Lagrangian for the Born-Infeld electro-dynamics,

$$\begin{aligned} L^{BI-ED} &= \sqrt{\det[F^{\mu\nu} - \alpha\eta^{\mu\nu}]} \\ &= \epsilon_{\mu\nu\lambda\rho}[F^{\mu\nu} - \alpha\eta^{\mu\nu}][F^{\lambda\rho} - \alpha\eta^{\lambda\rho}] \\ &= \mathrm{Pfaff}[F^{\mu\nu} - \alpha\eta^{\mu\nu}]. \end{aligned} \tag{4.30}$$

The Lagrangian (4.28) contains only one free parameter (l) which, as explained above, can always be absorbed in a redefinition of the vielbein. This action has a number of interesting classical features like simple field equations,

$$\epsilon_{a_1 a_2 \cdots a_{2n}} e^{a_2} \bar{R}^{a_3 a_4} \cdots \bar{R}^{a_{2n-1} a_{2n}} = 0. \tag{4.31}$$

These equations admit black hole solutions, and reasonable cosmological models [60, 61, 64]. The simplification comes about because the equations admit a unique maximally symmetric solution that corresponds to (anti-)de Sitter spacetime,

$$R^{ab} \pm \frac{1}{l^2} e^a e^b = 0, \tag{4.32}$$

in contrast with the situation if all the a_p's are arbitrary. For arbitrary a_p's, on the other hand, the field equations do not completely determine the components of R^{ab} and T^a because the nonlinearities of the equations give rise to degeneracies. The BI choice is in this respect the best behaved of the Lovelock family in even dimensions, since the degeneracies only occur for one value of the radius of curvature ($R^{ab} \pm \frac{1}{l^2} e^a e^b = 0$). At the same time, the BI action has the least number of algebraic constrains among the field equations required by consistency, and it is therefore the one with the simplest dynamical behavior [62].

In four dimensions, the BI Lagrangian is the standard Einstein-Hilbert action with cosmological constant, plus the Euler term as in (4.9) but with the Lovelock coefficients a_0, a_1, and a_2 chosen so that

$$L_4^{BI} = \kappa \epsilon_{abcd} \bar{R}^{ab} \bar{R}^{cd},$$
$$= \kappa \epsilon_{abcd} \left[R^{ab} R^{cd} \pm \frac{2}{l^2} R^{ab} e^c e^d + \frac{1}{l^4} e^a e^b e^c e^d \right]. \tag{4.33}$$

This expression requires adjusting only one parameter, which can be taken as the coefficient multiplying the Euler density. In principle the classical theory does not depend on the choice of that coefficient since the Euler density does not contribute to the field equations.

However, the Euler invariant is essential to render the action principle well defined for asymptotically (A)dS boundary conditions. In particular, the addition of the Euler density allows the regularization of the mass and angular momentum for the Kerr black hole in anti-de Sitter space [65]. The metric of asymptotically AdS geometry in standard Schwarzschild-like coordinates diverges at infinity, which means that the conserved charges which are generically given by boundary terms are extremely sensitive to the addition of a local function of the metric, as it can give a divergent contribution. For this reason, boundary terms in AdS spaces may dramatically change the charges even if the classical equations are not affected. This result of using the BI action is not merely a trick to render finite a potentially divergent charge, but is a prescription for a well-defined action principle, in which the mass, angular momentum and all thermodynamic quantities are regularized without the need for ad-hoc substractions or other assumptions about the background geometry.

In spite of all the nice features of the BI theory, the Lovelock coefficients are not protected by gauge symmetry. Although the maximally symmetric vacuum solution is invariant under both global and local AdS transformations, this is not a symmetry of the theory. It is a symmetry of one state only. In this sense, unlike the CS theories, the BI action is unlikely to be untouched by renormalization.

Chapter 5

Chern-Simons Gravities

5.1 Selecting Sensible Theories in Three Dimensions

The coefficients a_p in the Lovelock action (4.2) have dimensions l^{2p-D} to compensate for the canonical dimension of the vielbein ($[e^a] = l$) and the Lorentz connection ($[\omega^{ab}] = l^0$). The exterior derivative $d = dx^\mu \partial_\mu$ is dimensionless, and so is any genuine gauge connection like ω. The presence of dimensionful parameters leaves little room for optimism for a quantum version of the theory. Dimensionful parameters in the action are potentially dangerous because they are likely to acquire uncontrolled quantum corrections. This is what makes ordinary gravity nonrenormalizable in perturbation theory: In four dimensions, Newton's constant has dimensions of $[\text{mass}]^{-2}$ in natural units. It means that as the order in perturbation series increases, more powers of momentum will occur in the Feynman graphs, making the ultraviolet divergences increasingly worse. Concurrently, the radiative corrections to these bare parameters require the introduction of infinitely many counterterms into the action to render them finite [14]. But the illness that requires infinite amount of medication is obviously incurable.

The only safeguard against the threat of uncontrolled divergences in quantum theory is to have a symmetry principle that fixes the values of the parameters in the action, limiting the number of possible counterterms that could be added to the Lagrangian. Obviously, a symmetry endowed with such a high responsibility should be a bona fide quantum symmetry, and not just an approximate feature of its effective classical descendent. A symmetry that is only present in the classical theory but is not a feature of the quantum theory is said to be anomalous. This means that if one conceives the quantum theory as the result of successive quantum corrections to the classical theory, these corrections would "break" the symmetry. Of course, we know that the classical theory is a limit of the quantum world, some sort of shadow of an underlying reality that is blurred in the limit. An anomalous symmetry is an artifact of the classical limit, that does not correspond to a true symmetry of the microscopic world.

Thus, if a "non-anomalous" symmetry fixes the values of the parameters in the action, this symmetry will protect those values under renormalization. A good indication that this might happen would be if all the coupling constants are dimensionless and could be absorbed in the fields, as in four-dimensional Yang-Mills theory. As shown below, in odd dimensions there is a unique combination of terms in the action that can give the theory an enlarged gauge symmetry. The resulting action can be seen to depend on a unique multiplicative coefficient playing the role of Newton's constant. Moreover, this coefficient can be shown to be quantized [66] by an argument similar to the one that yields Dirac's quantization of the product of magnetic and electric charge [67].

5.1.1   *Extending the Lorentz group*

The presence of arbitrary dimensionful parameters is a problem that already occurs in three dimensions. Therefore it could be useful to analyze how to cure this problem in this case first. If we fail in the simplest example, the problem is likely to become intractable in general, but if one succeeds here, perhaps the lesson can be used to go into higher dimensions. Three-dimensions, in spite of being a very special situation for many different reasons, will provide some clues to extend the local Lorentz invariance to a bigger symmetry group in a natural way, as the Poincaré or the (anti)-de Sitter groups. This laboratory case will be our guiding principle to generalize the three-dimensional case to higher odd dimensions.

The dimensionality of the coefficients a_p in Lovelock's action (4.2) reflects that gravity is a gauge theory for the Lorentz group, where the vielbein e^a *is not* a connection but a vector under Lorentz rotations and plays the role of a matter field. If the vielbein is rescaled as $e^a \rightarrow \hat{e}^a := e^a/l$, the Lovelock coefficients are also rescaled $a_p \rightarrow \hat{a}_p = a_p/l^{2p-D}$ and therefore dimensionless. In this way one manages to hide the parameter l under the rug, but it will pop up as soon as one tries to describe the geometry in terms of the metric, because then a fundamental dimensionful scale is needed to do physics.

Three-dimensional gravity is an important exception to the previous statement. The Lagrangian in (4.2) reads

$$L_3 = \epsilon_{abc}(a_1 R^{ab} e^c + a_0 e^a e^b e^c), \qquad (5.1)$$

where a_0 and a_1 have dimensions [length]$^{-3}$ and [length]$^{-1}$, respectively. The exception is that by a suitable rescaling of the vielbein $e^a \rightarrow \hat{e}^a = e^a \sqrt{|\frac{3a_0}{a_1}|}$, (see (5.18) below), the coefficients a_0 and a_1 can be turned into a global dimensionless factor multiplying the action, and a fixed relative coefficient that can take the values $0, \pm\frac{1}{3}$. These values precisely make the corresponding action into a gauge theory for the

Poincaré, the anti-de Sitter or the de Sitter groups, respectively:

$$\textbf{i)}\ a_0 = 0,\ \text{then}\ L_3 \to L_3^{(0)} = \epsilon_{abc} R^{ab} \hat{e}^c, \tag{5.2}$$

$$\textbf{ii)}\ a_0 a_1 > 0,\ \text{then}\ L_3 \to L_3^{(+)} = \epsilon_{abc} \left(R^{ab} \hat{e}^c + \frac{1}{3} \hat{e}^a \hat{e}^b \hat{e}^c \right), \tag{5.3}$$

$$\textbf{iii)}\ a_0 a_1 < 0,\ \text{then}\ L_3 \to L_3^{(-)} = \epsilon_{abc} \left(R^{ab} \hat{e}^c - \frac{1}{3} \hat{e}^a \hat{e}^b \hat{e}^c \right). \tag{5.4}$$

As will become clear, these actions are not quite invariant under a gauge transformation, but they transform by a closed form, something that can be locally expressed as a total derivative. This *quasi-invariance* is enough to make the classical Euler-Lagrange equations invariant. Quasi-invariance is also sufficient to apply Noether's theorem to determine the conserved charge that generates the gauge symmetry of the action.

5.1.2 *Local Poincaré (quasi-) invariance*

The three-dimensional Einstein-Hilbert Lagrangian (5.2)

$$L_3 = \epsilon_{abc} R^{ab} e^c, \tag{5.5}$$

is Lorentz invariant: it is a contracted product of the tensors ϵ_{abc}, R^{ab} and e^c. This can be confirmed by performing an infinitesimal Lorentz transformation parameterized by $\lambda^a{}_b$, under which the vielbein and Lorentz connection transform as

$$\delta_{Lor} e^a = -\lambda^a{}_c e^c, \tag{5.6}$$

$$\delta_{Lor} \omega^a{}_b = D\lambda^a{}_b$$
$$= d\lambda^a{}_b + \omega^a{}_c \lambda^c{}_b - \omega^c{}_b \lambda^a{}_c. \tag{5.7}$$

From (5.7), R^{ab} can be seen to transform as a tensor, so that

$$\delta_{Lor} R^{ab} = \lambda^a{}_c R^{cb} + \lambda^b{}_c R^{ac}, \tag{5.8}$$

$$\delta_{Lor} \epsilon_{abc} = -(\lambda^d{}_a \epsilon_{dbc} + \lambda^d{}_b \epsilon_{adc} + \lambda^d{}_c \epsilon_{abd}) \equiv 0. \tag{5.9}$$

Combining these relations, it is direct to check that the Einstein-Hilbert Lagrangian (5.5) is invariant under infinitesimal Lorentz transformations, $\delta_{Lor} L_3 = 0$. Since the composition of two Lorentz transformations is also a Lorentz transformation, the argument can be iterated into a finite transformation.

It is unexpected that the action defined by (5.5) is also symmetric under the group of local translations in the tangent space. For this additional symmetry, e^a transforms as a gauge connection for the translations parameterized by $\xi^a(x)$ while the spin connection is unchanged,

$$\delta_{\text{Trans}} e^a = D\xi^a = d\xi^a + \omega^a{}_b \xi^b, \tag{5.10}$$

$$\delta_{\text{Trans}} \omega^{ab} = 0. \tag{5.11}$$

Then, the Lagrangian changes by a total derivative,

$$\delta_{\text{Trans}} L_3^{(0)} = d \left[\epsilon_{abc} R^{ab} \xi^c \right]. \tag{5.12}$$

This boundary term does not affect the classical field equations. Moreover, it vanishes under the standard assumption of asymptotically flat boundary conditions. This means that the three-dimensional GR action is invariant under all local Poincaré transformations.

It can be easily proved that transformations (5.6), (5.7), (5.10), and (5.11) form a representation of the Poincaré algebra. Consider two translations parametrized by ξ_1^a, ξ_2^a,

$$[\delta_{\xi_2}, \delta_{\xi_1}] \left\{ \begin{array}{c} e^a \\ \omega^a{}_b \end{array} \right\} = 0, \tag{5.13}$$

which translates into $[P_a, P_b] = 0$. For two Lorentz rotations $\lambda_1^{ab}, \lambda_2^{ab}$,

$$[\delta_{\lambda_2}, \delta_{\lambda_1}] \left\{ \begin{array}{c} e^a \\ \omega^a{}_b \end{array} \right\} = \delta_{[\lambda_2, \lambda_1]} \left\{ \begin{array}{c} e^a \\ \omega^a{}_b \end{array} \right\}, \tag{5.14}$$

where $[\lambda_1, \lambda_2]^a{}_b = (\lambda_1)^a{}_c (\lambda_2)^c{}_b - (\lambda_2)^a{}_c (\lambda_1)^c{}_b$, which translates into $[J_{ab}, J_{cd}] = \eta_{bc} J_{ad} - \eta_{ac} J_{bd} + \eta_{ad} J_{bc} - \eta_{bd} J_{ac}$. Finally,

$$[\delta_\lambda, \delta_\xi] \left\{ \begin{array}{c} e^a \\ \omega^a{}_b \end{array} \right\} = \delta_{[\lambda, \xi]} \left\{ \begin{array}{c} e^a \\ \omega^a{}_b \end{array} \right\}, \tag{5.15}$$

where $[\lambda, \xi]^a = \lambda^a{}_c \xi^c$, which translates into $[J_{ab}, P_c] = \eta_{bc} P_a - \eta_{ac} P_b$. Thus, the local Poincaré transformations acting on the vielbein and the Lorentz connection is replicated by the Poincaré algebra. Thus, the invariance of the Lagrangian (5.12) has been extended to a Poincaré symmetry of the theory.

5.1.3 *Local (anti-)de Sitter symmetry*

In the presence of a cosmological constant, $\Lambda \neq 0$, it is also possible to extend the local Lorentz symmetry to the local de Sitter ($\Lambda > 0$), or anti-de Sitter ($\Lambda < 0$) groups. The point is that different spaces $T^* M$ can be chosen as tangents to a given manifold M, provided they are diffeomorphic to the open neighborhoods of M.

Useful choices of tangent spaces are the covering space of a vacuum solution of the Einstein equations and maximally symmetric spaces. In the previous case, flat space was singled out because it is the maximally symmetric solution of the Einstein equations for $\Lambda = 0$, while if $\Lambda \neq 0$, flat spacetime is no longer a solution of the Einstein equations and thus de Sitter and anti-de Sitter spaces are now the best options. For nonvanishing cosmological constant, the Lagrangian

$$L_3 = \frac{1}{2\kappa} \int \sqrt{-g} \left[R - 2\Lambda \right] d^3 x, \tag{5.16}$$

can be compared with (4.7)

$$L_3 = \epsilon_{abc}(a_1 R^{ab} e^c + a_0 e^a e^b e^c).$$ (5.17)

Therefore, $a_0 = \Lambda(6\kappa)^{-1}$ and $a_1 = -(4\kappa)^{-1}$ and consequently $a_0 a_1 > 0$ corresponds to $\Lambda < 0$ (anti-de Sitter), and vice-versa.

We will show that this Lagrangian changes by a boundary term under gauge transformations for an extension of the three-dimensional Lorentz group, the de Sitter or anti-de Sitter groups $SO(3,1)$ or $SO(2,2)$, respectively. The essence of the proof is the observation that ω^{ab} and e^a can be combined as a connection for the (a)dS group. As a preliminary step, let us observe that the vielbein can be rescaled as

$$e^a \rightarrow \hat{e}^a = \sqrt{\left|\frac{3a_0}{a_1}\right|} e^a,$$ (5.18)

and the Lagrangian (5.17) becomes

$$L_3 = k\epsilon_{abc}\left(R^{ab}\hat{e}^c - \sigma\frac{1}{3}\hat{e}^a\hat{e}^b\hat{e}^c\right),$$ (5.19)

where $k = \sqrt{|a_1^3/3a_0|}$ is a dimensionless quantity and $\sigma = -\text{sign}(a_0 a_1)$. The new vielbein $\hat{e}^a$ is a dimensionless one-form, just like the connection, and therefore one can consider the combined one-form

$$W^{AB} = \begin{bmatrix} \omega^{ab} & \hat{e}^a \\ -\hat{e}^b & 0 \end{bmatrix},$$ (5.20)

where $a, b, .. = 0, 1, 2$ and $A, B, ... = 0, 1, 2, 3$. This new one-form is antisymmetric and hence it can also be viewed as connection for an extension of the Lorentz group, $G \supseteq SO(2,1)$, which must have six generators as this is the number of components of the connection W^{AB}. Under a local G-transformation, one expects W^{AB} to transform in the same way that ω transforms under $SO(2,1)$ in (5.7),

$$\delta W^{AB} = d\lambda^{AB} + W^A{}_C\lambda^{CB} + W^B{}_C\lambda^{AC} =: D^{(W)}\lambda^{AB},$$ (5.21)

where $\Lambda^A{}_B = \delta^A{}_B + \lambda^A{}_B$ is an element of G infinitesimally close to the identity and $D^{(W)}$ stands for the covariant derivative in the connection W. The infinitesimal elements in λ can be split in to those in $SO(2,1)$ and the rest,

$$\lambda^{AB} = \begin{bmatrix} \lambda^{ab} & \hat{\xi}^a \\ -\hat{\xi}^b & 0 \end{bmatrix},$$ (5.22)

In order to make sense of (5.21) one should indicate how the indices A, B are to be lowered. In the case of ω^{ab}, the indices were raised and lowered using invariant metric of the three-dimensional Lorentz group, η_{ab}. In the new situation we may use

$$\Pi_{AB} = \begin{bmatrix} \eta_{ab} & 0 \\ 0 & \sigma \end{bmatrix},$$ (5.23)

where $\sigma = -1$ corresponds to $G = SO(2,2)$ while $\sigma = +1$ corresponds to $G = SO(3,1)$. Additionally, the Levi-Civita invariant tensor ϵ_{ABCD} can be chosen so that

$$\epsilon_{abc3} = \epsilon_{abc}. \tag{5.24}$$

The connection W defines a curvature two-form, given by

$$F^{AB} = dW^{AB} + W^A{}_C W^{CB} = \begin{bmatrix} R^{ab} - \sigma \hat{e}^a \hat{e}^b & \hat{T}^a \\ -\hat{T}^b & 0 \end{bmatrix}, \tag{5.25}$$

where $\hat{T}^a = d\hat{e}^a + \omega^a{}_b \hat{e}^b$. With this curvature and (5.24), the Euler density can be defined,

$$\begin{aligned} \mathfrak{E}_4 &= \epsilon_{ABCD} F^{AB} F^{CD} \\ &= 4\epsilon_{abc}(R^{ab} - \sigma \hat{e}^a \hat{e}^b)\hat{T}^c, \end{aligned} \tag{5.26}$$

which can be expressed as an exterior derivative,

$$\begin{aligned} \mathfrak{E}_4 &= 4\epsilon_{abc}\left(D[R^{ab}\hat{e}^c] - \frac{\sigma}{3}D[\hat{e}^a\hat{e}^b\hat{c}^c]\right) \\ &= 4d\left(\epsilon_{abc}\left[R^{ab}\hat{e}^c - \frac{\sigma}{3}\hat{e}^a\hat{e}^b\hat{e}^c)\right]\right). \end{aligned} \tag{5.27}$$

The expression in parenthesis is precisely (5.19), and hence we conclude that the exterior derivative of the 3D gravitational Lagrangian with cosmological constant is the 4D Euler density which is, by construction, invariant under $G = SO(2,2)$ or $G = SO(3,1)$. The importance of this result is that the Lagrangian (5.19) inherits this enlarged symmetry, because

$$\delta_G \mathfrak{E}_4 = 0 \Rightarrow \delta_G(d\hat{L}_3) = d(\delta_G \hat{L}_3) = 0, \tag{5.28}$$

which means that at least in an open patch $\delta_G \hat{L}_3 = d(\text{something})$. In other words, under the enlarged gauge group the 3D Lagrangian transforms by a total derivative. This can also be explicitly confirmed by writing (5.21) in terms of the variations of ω^{ab} and $\hat{e}^a$,

$$\delta\omega^{ab} = d\lambda^{ab} + \omega^a{}_c\lambda^{cb} + \omega^b{}_c\lambda^{ac} \pm [\hat{e}^a\hat{\xi}^b - \hat{\xi}^a\hat{e}^b] \tag{5.29}$$

$$\delta\hat{e}^a = d\hat{\xi}^a + \omega^a{}_b\hat{\xi}^b - \lambda^a{}_b\hat{e}^b, \tag{5.30}$$

which we leave as an exercise for the interested readers.

Another simple consistency check is that the Poincaré symmetry is obtained in the limit $l \to \infty$ ($\lambda \to 0$) and the transformations (5.29, 5.30) approach those in (5.6, 5.7, 5.10, 5.11), where $\hat{e}^a = e^a/l$ and $\hat{\xi}^a := \xi^a/l$. The vanishing cosmological constant limit is actually a deformation of the (A)dS algebra analogous to the deformation that yields the Galileo group from the Poincaré symmetry in the limit of infinite speed of light $c \to \infty$. These deformations are examples of what is known as an Inönü-Wigner contraction [68, 69] which we will discuss later. The procedure starts from a semisimple Lie algebra and some generators are rescaled by a parameter ($\sqrt{|\Lambda|}$ or l in the above example). Then, in the limit $\Lambda \to 0$ (or $l \to \infty$), a new non-semisimple algebra of the same dimension is obtained. For the Poincaré group which is the familiar symmetry of Minkowski space, the representation in terms of W becomes inadequate because the metric Π^{AB} should be replaced by the degenerate (non-invertible) metric of the Poincaré group,

$$\Pi_0^{AB} = \begin{bmatrix} \eta^{ab} & 0 \\ 0 & 0 \end{bmatrix}, \tag{5.31}$$

and is no longer clear how to raise and lower indices. Nevertheless, the Lagrangian (5.17) in the limit $l \to \infty$ takes the usual Einstein-Hilbert form with vanishing cosmological constant (5.5).

There is still an issue that might bother the attentive reader. The Lagrangian (5.19) is a three-form in a three-dimensional manifold M^3, therefore its exterior derivative is necessarily vanishing. A rigorous way to present this is to conceive the three-dimensional manifold as embedded in a higher dimensional space where L_3 is actually defined and therefore dL_3 need not vanish. Another way is to imagine the manifold $\overline{M}$ where the theory is defined is actually four-dimensional and M^3 is actually a boundary $M^3 = \partial \overline{M}$.

5.1.4 *Three-dimensional zoo*

Up to now, we have only considered the Euler characteristic, but this is not the only invariant four-form available. As mentioned in Section 4.3, there are two other families of invariants, the Pontryagin forms (4.26) and, if torsion is also included, the Nieh-Yan invariants (4.22). For each of these four-forms there is a locally defined 3−form whose exterior derivative yields the corresponding closed form as

$$\mathcal{C}_3^{\text{Pont}} = \omega^a{}_b d\omega^b{}_a + \frac{2}{3} \omega^a{}_b \omega^b{}_c \omega^c{}_a, \quad d\mathcal{C}_3^{\text{Pont}} = R^{ab} R_{ba}, \tag{5.32}$$

$$\mathcal{C}_3^{NY} = e^a T_a, \quad d\mathcal{C}_3^{NY} = T^a T_a - e^a e^b R_{ab}. \tag{5.33}$$

These forms, which are invariant under $SO(2,1)$, can also be included as Lagrangian densities in three dimensions. However, it is interesting to note that there is a particular combination of them that is quasi-invariant under the full (anti-) de Sitter group, sometimes called *exotic Lagrangian* in the literature [10] given by

$$L_3^{\text{Exotic}} = \mathcal{C}_3^{\text{Pont}} \pm \frac{2}{l^2} \mathcal{C}_3^{NY} = \omega^a{}_b d\omega^b{}_a + \frac{2}{3} \omega^a{}_b \omega^b{}_c \omega^c{}_a \pm \frac{2}{l^2} e_a T^a. \tag{5.34}$$

This is the Chern-Simons form associated to the Pontryagin form for the (anti-) de Sitter curvature,

$$dL_3^{\text{Exotic}} = F^A{}_B F^B{}_A = \mathfrak{P}_4, \tag{5.35}$$

where the indices are raised and lowered with (5.23). This Lagrangian has the curious property of giving exactly the same field equations as the standard L_3^{AdS}, but interchanged: varying with respect to e^a one gives the equation for ω^{ab} of the other, and vice-versa.

Let us summarize all the possible first order gravitational Lagrangians in $D = 3$ and their relations with the four-dimensional characteristic forms. We voluntarily omit the hat in the expressions of the vielbein

Table 5.2

Lagrangian	Characteristic form	Group
$L_3^{(A)dS} = \varepsilon_{abc}(R^{ab} \pm \frac{1}{3}e^a e^b)e^c$	$\mathfrak{E}_4 = \epsilon_{ABCD}F^{AB}F^{CD}$	$SO(2,2)(+)$ $SO(3,1)(-)$
$L_3^{\text{Poincaré}} = \epsilon_{abc}R^{ab}e^c$	—	$SO(2,1)$
$L_3^{\text{Pont}} = \omega^a{}_b d\omega^b{}_a + \frac{2}{3}\omega^a{}_b \omega^b{}_c \omega^c{}_a$	$\mathfrak{P}_4^{\text{Pont}} = R^a{}_b R^b{}_a$	$SO(2,1)$
$L_3^{NY} = e^a T_a$	$\mathfrak{N}_4^{NY} = T^a T_a - e^a e^b R_{ab}$	$SO(2,1)$
$L_3^{\text{Exotic}} = L_3^{\text{Pont}} \pm 2L_3^{NY}$	$\mathfrak{P}_4^{(A)dS} = F^A{}_B F^B{}_A$	$SO(2,2)(+)$ $SO(3,1)(-)$

Here R and F are the curvatures of the Lorentz and (anti-) de Sitter connections $\omega^a{}_b$, $W^A{}_B$, respectively; T is the torsion; $\mathfrak{E}_4$, $\mathfrak{P}_4$ and $\mathfrak{N}_4$ are the Euler, Pontryagin and Nieh-Yan invariants. The Lagrangians are quasi-invariant (they change by a total derivative) under the corresponding gauge groups. Note that the standard gravitational Lagrangians in the presence of cosmological constant $L_3^{(A)dS}$ are CS forms: their exterior derivative are characteristic forms whose integral is a topological invariant. Their vanishing cosmological constant limit $L_3^{\text{Poincaré}}$, however is not a CS form in this sense, since its exterior derivative,

$$dL_3^{\text{Poincaré}} = \epsilon_{abc}R^{ab}T^c, \tag{5.36}$$

is not a characteristic form in four dimensions, the three index Levi-Civita tensor is not an invariant of either $SO(4)$, $SO(3,1)$ or $SO(2,2)$.

5.2 More Dimensions

What has been said about embedding the Lorentz group into the Poincaré or (a)dS groups for $D = 3$, can be generalized for other dimensions. In fact, it is always possible to embed the D-dimensional Lorentz group into the de-Sitter, or anti-de Sitter groups,

$$SO(D-1,1) \hookrightarrow \begin{cases} SO(D,1), & \Pi^{AB} = \text{diag}(\eta^{ab}, +1) \\ SO(D-1,2), & \Pi^{AB} = \text{diag}(\eta^{ab}, -1) \end{cases}. \tag{5.37}$$

and in the Poincaré limit ($\Lambda = 0$),

$$SO(D-1,1) \hookrightarrow ISO(D-1,1). \tag{5.38}$$

The question is whether the Lorentz symmetry for the gravity actions can also be extended to an $SO(D,1)$, $SO(D-1,2)$, $ISO(D-1,1)$ symmetry in dimensions ≥ 3.

As we will see, the answer to this question is affirmative in odd dimensions: There exist gravity actions for every $D = 2n - 1$, quasi-invariant under local $SO(2n-2, 2)$, $SO(2n-1, 1)$ or $ISO(2n-2, 1)$ transformations. In these theories the vielbein and the spin connection combine to form the connection of the larger group. In even dimensions, however, this cannot be done.

In order to understand this, let us examine how the trick works for D = 3. Why was it possible to enlarge the symmetry from local $SO(2,1)$ to local $SO(3,1)$, $SO(2,2)$ and $ISO(2,1)$ in this case? What happens if one tries to do this for $D \geq 4$?

5.2.1 *The lesson from D = 3*

Let us start with the Poincaré group and the Einstein-Hilbert action in four dimensions,

$$L_4 = \epsilon_{abcd} R^{ab} e^c e^d. \tag{5.39}$$

Under local translations $\delta e^a = d\lambda^a + \omega^a{}_b \lambda^b$ a simple calculation yields

$$\begin{aligned} \delta L_4 &= 2\epsilon_{abcd} R^{ab} e^c \delta e^d \\ &= d(2\epsilon_{abcd} R^{ab} e^c \lambda^d) - 2\epsilon_{abcd} R^{ab} T^c \lambda^d. \end{aligned} \tag{5.40}$$

The first term in the r.h.s. of (5.40) is a total derivative and therefore gives a surface contribution to the action that can be ignored. The last term, however, need not vanish unless one imposes $T^a = 0$. This condition is a consequence of the equation of motion $\delta L/\delta\omega = 0$, so it might be argued to be ok: the invariance of the action under local translations is an *on shell symmetry*. However, "on shell symmetries" are not real symmetries and they are unlikely to survive quantization because quantum mechanics doesn't respect equations of motion.[1]

The miracle in three dimensions occurred because the Lagrangian (5.5) is linear in the vielbein e. In fact, Lagrangians of the form

$$L_{2n+1} = \epsilon_{a_1 \cdots a_{2n+1}} R^{a_1 a_2} \cdots R^{a_{2n-1} a_{2n}} e^{a_{2n+1}}, \tag{5.41}$$

which are only defined in odd dimensions, are also invariant under local Poincaré transformations (5.10) and (5.11), as can be easily checked. Since the Poincaré group is a limit of (A)dS, it seems likely that there should exist a Lagrangian in odd dimensions, invariant under local (A)dS transformations, whose limit for vanishing cosmological constant $(l \to \infty)$ is (5.41).

---

[1] The symmetry under local translations (5.40) could also result from $R^{ab} = 0$, but this is not even one of the field equations (remember that R^{ab} is the curvature two-form, not related to the Ricci tensor $R^{\mu\nu} = g^{\alpha\beta} R^{\mu\nu}{}_{\alpha\beta}$).

One way to find out what the (A)dS symmetric theory might be, one could take the most general Lovelock Lagrangian and select the coefficients by requiring invariance under (5.29) and (5.30). This is a long and tedious but sure route. An alternative approach is to try to understand why it is that in three dimensions the gravitational action with cosmological constant (5.17) is symmetric under local (A)dS transformations.

It is a unique feature of three-dimensional gravity that the Lovelock action possesses a larger symmetry than could have been naively anticipated for all values of a_0 and $a_1 \neq 0$. The obvious Lorentz invariance of the Lagrangian (5.17) is enhanced in the three cases $a_0 a_1 = 0$, < 0 or > 0, into the Poincaré, the de Sitter or the anti-de Sitter groups, respectively. This was noted by Achúcarro and Townsend [70] and its quantum consequences have been extensively discussed by Witten [71, 72].

The enhancement of symmetry does not result from a modification or a fine tuning of the parameters in the action, but from the observation that the fields ω and e can be considered independent parts of a gauge connection for the larger group. The resulting action is not invariant under the larger group, but *quasi-invariant*, it changes by a total derivative. The reason it changes by a closed form is that the Lagrangian is such that its exterior derivative is gauge invariant (characteristic form) under the larger symmetry group. Objects whose exterior derivative are characteristic forms are known as *Chern-Simons forms* and were discovered almost by accident in mathematics [73], but had been used in physics in mechanics and in electrodynamics for the past two centuries [74].

Let us make this more explicit. A *Characteristic 2n-form*, also known as a *characteristic class* is the product of n curvatures,

$$\mathcal{Q}_{2n}(A) = \langle F \wedge F \wedge \cdots \wedge F \rangle, \tag{5.42}$$

where $F = dA + A^2$, A is a gauge connection in the Lie algebra of some gauge group and $\langle \cdots \rangle$ is a symmetrized trace over the internal gauge indices. Then $\mathcal{Q}_{2n}$ is invariant under gauge transformations, $A \to A' = g^{-1}(d + A)g$, and by virtue of the Bianchi/Jacobi identity $DF = 0$ it can be shown that $d\mathcal{Q}_{2n} \equiv 0$ (see, e.g., [8]). Therefore by Poincaré's Lemma $\mathcal{Q}_{2n}(A)$ can be locally expressed as a total derivative,[2]

$$\mathcal{Q}_{2n}(A) = d\mathcal{C}_{2n-1}(A). \tag{5.43}$$

The $(2n - 1)$-form $\mathcal{C}_{2n-1}(A)$ is a Chern-Simons (**CS**) form. Since the derivative of a CS form is invariant, the CS form itself inherits a weaker form of invariance changing by a total derivative, it is quasi-invariant.

In particular, the Lagrangian (5.19) is the Chern-Simons form associated with the four-dimensional Euler form for the (anti-) de Sitter group. What we have

---

[2]Poincaré's Lemma states that if in an open simply connected region a p-form Ω is closed ($d\Omega = 0$), then it is always possible to write it in a local patch as the exterior derivative of a $(p-1)$-form [40].

found here is an explicit way to immerse the three-dimensional Lorentz group into a larger symmetry group, in which both, the vielbein and the Lorentz connection have been incorporated as independent components of a larger (A)dS connection. Written in this fashion, the CS nature of the Lagrangian and its quasi-invariance become apparent.

The idea of characteristic class is one of the unifying concepts in mathematics that connects differential geometry, algebraic topology and, through the index theorems, to the spectra of differential operators. The theory of characteristic classes explains mathematically why it is not always possible to perform a gauge transformation that makes the connection vanish everywhere even if it is locally pure gauge. The non-vanishing value of a topological invariant signals an obstruction to the existence of a gauge transformation that trivializes the connection globally. This obstruction can be traced to the CS form integrated over the boundary of a compact manifold and it is intriguing that these CS forms can also provide an action principle for the geometry of an odd-dimensional spacetime.

We learned in the three-dimensional case that if one takes seriously the idea that W^{AB} in (5.20) is a connection, then its curvature $F^{AB} = dW^{AB} + W^A_C W^{CB}$ can be used to define characteristic forms in four dimensions

$$\mathfrak{E}_4 = \epsilon_{ABCD} F^{AB} F^{CD}, \quad \mathfrak{P}_4 = F^A{}_B F^B{}_A. \tag{5.44}$$

These forms are invariant under the (A)dS group and can be locally expressed as the exterior derivative of CS three forms, that define three-dimensional Lagrangians.

5.2.2 *Lovelock-Chern-Simons theories*

Consider the Euler density in four dimensions which can also be written explicitly in terms of R^{ab}, T^a, and e^a,

$$\mathfrak{E}_4 = 4\epsilon_{abc}(R^{ab} \pm l^{-2} e^a e^b) l^{-1} T^a$$

$$= 4 l^{-1} d \left[\epsilon_{abc} \left(R^{ab} \pm \frac{1}{3l^2} e^a e^b \right) e^c \right]$$

$$= \# d L_3^{(A)dS}. \tag{5.45}$$

This is the key point: Since $\mathfrak{E}_4$ is invariant under local (A)dS$_3$, the variation of the right hand side under a local (A)dS transformation must vanish, $\delta(dL_3^{(A)dS}) = 0$. The variation δ is a linear operation, therefore $d(\delta L_3^{(A)dS}) = 0$, which in turn means that, locally, $\delta L_3^{(A)dS} = d(\text{something})$ and the action is (A)dS invariant up to surface terms.

The procedure to generalize the (A)dS Lagrangian from 3 to $2n-1$ dimensions is now clear [60, 61, 75, 76]:

• First, observe that the expression of the connection W^{AB} for the extended group can be written in any dimension D.

- Then, for $D + 1 = 2n$, the Euler-form analogous to (5.26) is,

$$\mathfrak{E}_{2n} = \epsilon_{A_1 \cdots A_{2n}} F^{A_1 A_2} \cdots F^{A_{2n-1} A_{2n}}. \tag{5.46}$$

- Using (5.25) $\mathfrak{E}_{2n}$ can be expressed in terms of R^{ab}, T^a and e^a.
- Write this as the exterior derivative of a $(2n - 1)$-form L_{2n-1}.
- L_{2n-1} can be used as a Lagrangian in $D = 2n - 1$ dimensions and is (A)dS invariant by construction.

Proceeding in this way, the $(2n - 1)$-dimensional (A)dS invariant Lagrangian as

$$L_{2n-1}^{(A)dS} = \sum_{p=0}^{n-1} \bar{a}_p L^{(2n-1,p)}, \tag{5.47}$$

where $L^{(D,p)}$ is given by (4.3). This is a particular case of a Lovelock Lagrangian in which all the coefficients $\bar{a}_p$ have been fixed as

$$\bar{a}_p = \kappa \cdot \frac{(\pm 1)^{p+1} l^{2p-D}}{(D - 2p)} \binom{n-1}{p}, \tag{5.48}$$

where $1 \le p \le n - 1 = (D - 1)/2$, κ is an arbitrary dimensionless constant and the signs correspond to anti-de Sitter $(+)$ and de Sitter $(-)$. It is left as an exercise to the reader to check that $dL_{2n-1}^{(A)dS} = \mathfrak{E}_{2n}$ and the quasi-invariance of $L_{2n-1}^{(A)dS}$ under the (A)dS group. Another interesting exercise is to show that the action (5.47) can also be written as

$$I_{2n-1} = \frac{\kappa}{l} \int\limits_M \int\limits_0^1 dt\, \epsilon_{a_1 \cdots a_{2n-1}} R_t^{a_1 a_2} \cdots R_t^{a_{2n-3} a_{2n-2}} e^{a_{2n-1}}, \tag{5.49}$$

where we have defined $R_t^{ab} := R^{ab} \pm (t^2/l^2) e^a e^b$ [77].

5.2.3   *Locally Poincaré invariant theory*

If one blindly takes the limit $l \to \infty$ in the above expression, nothing survives. The next meaningful contribution in the limit of large l comes from the term linear in the vielbein. This is precisely the Poincaré invariant Lagrangian defined as

$$L_P = \kappa \epsilon_{a_1 \cdots a_{2n+1}} R^{a_1 a_2} \cdots R^{a_{2n-1} a_{2n}} e^{a_{2n+1}}, \tag{5.50}$$

which in addition to being invariant under local Lorentz transformations, changes by a total derivative under local translations. The Lagrangian L_P defines a Chern-Simons form of the Poincaré group

$$dL_P = \langle F^{n+1} \rangle,$$

where the invariant $(n + 1)$-tensor on the Poincaré group is given by

$$\langle J_{a_1 a_2}, \ldots, J_{a_{2n-1} a_{2n}}, P_{a_{2n+1}} \rangle = \epsilon_{a_1 \cdots a_{2n+1}}, \tag{5.51}$$

provides the invariant trace for the CS construction.

5.3   Torsional Chern-Simons Gravities

As will be seen in Chapter 8, another exceptional feature of CS gravities is the existence of natural supersymmetric extensions obtained by extending the (A)dS algebra to a superalgebra. In the construction of the supersymmetric extensions of the Lovelock-CS theories we will find that other CS forms, not related to the Euler but to the Pontryagin class that involve torsion explicitly, are also needed.

As in the three-dimensional case, in order to explicitly include torsional terms in the gauge Lagrangian, one must not use the completely antisymmetric tensor $\epsilon_{A_1 \cdots A_{2n}}$. For example, in four dimensions the Pontryagin invariant form (5.35)

$$\mathfrak{P}_4 = F^A{}_B F^B{}_A = dL_3^{\text{Exotic}}, \tag{5.52}$$

provides the three-dimensional *exotic Lagrangian* L_3^{exotic} (5.34). Note that since $F^A{}_B$ involves the torsion, so does L_3^{Exotic}.

The pattern to repeat this construction in higher dimension must take into account the antisymmetry of the curvature two-form F^{AB} are in A-B, and hence the trace defined previously must involve an even number of factors as

$$\mathfrak{P}_{4p} = F^{A_1}{}_{A_2} F^{A_2}{}_{A_3} \cdots F^{A_{2p}}{}_{A_1}. \tag{5.53}$$

Note that Pontryagin forms can be multiplied to produce other invariants,

$$\mathfrak{P}_{4p_1, 4p_2, \cdots 4p_k} = \mathfrak{P}_{4p_1} \mathfrak{P}_{4p_2} \cdots \mathfrak{P}_{4p_k}. \tag{5.54}$$

Invariant tensors like these can also be expressed locally as the exterior derivative of a $(4P - 1)$-form (with $P = p_1 + p_2 + \cdots + p_k$) and therefore these CS forms can also be used as Lagrangians in $(4P - 1)$ dimensions,

$$dL_T^{(A)dS} = \text{Tr}(F^{p_1}) \cdots \text{Tr}(F^{p_k}) \tag{5.55}$$

As an illustration, in seven dimensions there are three Lorentz-invariant torsional CS terms,

Table 5.3

$D = 7$ Torsional Chern-Simons Lagrangian	$\mathfrak{P}$
$L_7^{\text{Lorentz}} = \omega(d\omega)^3 + \cdots + \frac{4}{7}\omega^7$	$R^a{}_b R^b{}_c R^c{}_d R^d{}_a$
$L_7^A = (L_3^{\text{Lorentz}}) R^a{}_b R^b{}_a = \left(\omega^a{}_b d\omega^b{}_a + \frac{2}{3}\omega^a{}_b \omega^b{}_c \omega^c{}_a \right) R^a{}_b R^b{}_a$	$(R^a{}_b R^b{}_a)^2$
$L_7^B = (L_3^{\text{Torsion}}) R^a{}_b R^b{}_a = (e^a T_a)(R^a{}_b R^b{}_a)$	$(T^a T_a - e^a e^b R_{ab}) R^d{}_c R^c{}_d$

In addition, there are two combinations of torsional CS Lagrangians that have (A)dS$_7$ symmetry, corresponding to the two Pontryagin (A)dS 8-forms

$$\mathfrak{P}_8 = F^A{}_B F^B{}_C F^C{}_D F^D{}_A; , \quad \mathfrak{P}_{4,4} = (F^A{}_B F^B{}_A)(F^C{}_D F^D{}_C). \tag{5.56}$$

The corresponding CS 7-forms are

$$\mathcal{C}_7 = L_7^{\text{Lorentz}} + (T^a T_a + R^{ab} e_a e_b) T^c e_c + 4 T_a R_b^a R_c^b e^c$$

$$\mathcal{C}_{3,4} = \left(R^{ab} R_{ba} + 2(T^a T_a - R^{ab} e_a e_b) \right) L_3^{\text{Exotic}} \qquad (5.57)$$

where L_3^{Exotic} is defined in (5.34).

The CS construction is not restricted to the Euler invariant, but applies to all gauge invariant $2n$-forms of similar nature and for any Lie group, that is all characteristic classes [8]. Well known characteristic classes are the Pontryagin or Chern four-forms studied in the context of gauge theories, $\text{Tr}[F^2]$, so the corresponding CS forms

$$\text{Tr}\left[A dA + \frac{2}{3} A^3 \right], \qquad (5.58)$$

define useful Lagrangians in three dimensions. These are interesting mathematical models and have found many applications in condensed matter systems like high-Tc superconductivity [78–81], the quantum Hall effect [82–84], topological insulators [85] and graphene [86–88].

Chapter 6

Additional Features
of Chern-Simons Gravity

6.1 Lovelock-Chern-Simons Coefficients

The CS form has other properties that are potentially important for a quantum field theory. One remarkable feature is that all the Lovelock coefficients are fixed, dimensionless, rational numbers. As discussed in Section 4.4, the coefficient in the Lagrangian can be made dimensionless by absorbing a unit length in the definition of the vielbein, $\hat{e}^a = e^a/l$. The coefficients in the CS forms are rational numbers, because the characteristic form $\langle F^n \rangle$ is a polynomial in R^{ab}, T^a and e^a with integer coefficients, so "its integral" $\mathcal{C}_{2n-1}$ is a polynomial with rational coefficients.

The only free parameter in a Chern-Simons action is the overall factor κ. However, this coefficient can be seen to take discrete values in a quantum version of the theory. Consider a CS action for a simply connected, compact $2n - 1$ dimensional manifold M, which is the boundary of a $2n$-dimensional compact orientable manifold Ω. Then the action for the geometry of M can be expressed as the integral of the Euler density $\mathfrak{E}_{2n}$ over Ω, multiplied by κ. But since there can be many different manifolds with the same boundary M, the integral over Ω should give the same physical predictions as that over a different manifold, Ω' with the same boundary M. In order for this change to leave the path integral unchanged, a minimal requirement would be

$$\kappa \left[\int_\Omega \mathfrak{E}_{2n} - \int_{\Omega'} \mathfrak{E}_{2n} \right] = 2n\pi\hbar. \tag{6.1}$$

The quantity in brackets, with the appropriate normalization, is the Euler number of the manifold obtained by gluing Ω and Ω' along the common boundary[1] M, $\chi[\Omega \cup \Omega']$. This integral can take an arbitrary integer value and from this follows that κ must be quantized [66],

$$\kappa = nh, \tag{6.2}$$

where h is Planck's constant.

---

[1] This must be done in the right way to produce an orientable manifold.

Since all coefficients in the action-except for κ- are fixed by the enhanced (A)dS symmetry, they are protected from renormalization unless the gauge symmetry is somehow broken, which makes CS theories interesting candidates for quantum theories of gravitation in odd-dimensional spacetimes.

6.2   Stability of CS-Gravitation Theories

Lovelock-CS gravities have other nice features that make them exceptional among the family of Lovelock theories.

Generic Lovelock theory can have as many as $[(D-1)/2]$ different cosmological constants depending on as many arbitrary coefficients in the Lagrangian, which makes the "cosmological constant problem" much worse with increasing dimension [61]. This is also a potential source of ambiguities and instabilities of Lovelock gravities. For $D \geq 5$, there are classical solutions in which the cosmological constant could have different values in different regions and even jump arbitrarily between geometries with different Λs as the system evolves in time [50, 90]. Several cosmological constants also mean several possible event horizon radii for spherically symmetric black holes, and the geometry can jump chaotically between the different possibilities. In the generic case in five or more dimensions the vacuum has a ghost mode that signals quantum instability [89].

The Lovelock-Chern-Simons theories instead describe geometries with a unique cosmological constant and none of the above problems arise. There are other choices of the Lovelock coefficients that also lead to constant curvature solutions of a unique value and that admit static black holes with a unique horizon radius [64]. As shown in this reference, the CS theories are a particular exceptional subclass, their black hole solutions are completely different from those obtained with any other choice of Lovelock coefficients. CS black holes are the only ones that have positive specific heat for any mass and can reach thermal equilibrium with a thermal bath of any temperature. In contrast, Schwarzschild and Schwarzschild-AdS black holes of very small mass in all other Lovelock theories can never reach thermal equilibrium with a thermal bath and exhibit an explosive evaporation. This points to a fundamental instability in all Lovelock theories, with the only exception of the CS case.

6.3   Degrees of Freedom

A generic D-dimensional Lovelock Lagrangian has as many degrees of freedom as the Einstein-Hilbert Lagrangian in that dimension, i.e., $D(D-3)/2$ [90], because Lovelock theories in the torsion-free sector have the same independent fields as gravity in the metric formulation. Since the only symmetry of Lovelock — in the metric formulation — and Einstein-Hilbert gravity is spacetime diffeomorphisms, both theories have the same phase space coordinates and the same number of first class constraints.

In three dimensions, the Einstein-Hilbert theory has no local degrees of freedom or gravitational waves. This is true for the theory either in the metric or first order version, and with or without torsion. The matching of degrees of freedom between the generic Lovelock and Lovelock-CS breaks down at dimensions 5, and above. For $D = 2n + 1 \geq 5$, the Lovelock CS theories with $(D^2 + D - 4)/2$ propagating degrees of freedom [17], which is $2(D - 1)$ extra propagating degrees of freedom, are not contained in the generic Lovelock or Einstein-Hilbert theories.

6.4   Summary of Gravity Actions

The Lovelock action (4.2) is the most general action for gravity that does not involve torsion and gives up to second order field equations for the metric. This action has local Lorentz symmetry by construction, that can be extended in odd dimensions by an appropriate choice of the constants a_p to an (anti) de Sitter or Poincaré symmetry. In even dimensions, the Born-Infeld choice is a special case, but the symmetry is not enhanced. The following table summarized the different cases in the Lovelock family.

Table 6.1

Theory	Coefficients a_p	Dimension	Local symmetry
Generic	Arbitrary	Any D	$SO(D-1,1)$
Born-Infeld	$(\pm l)^{2p-D} \dbinom{n}{p}$	$D = 2n$	$SO(D-1,1)$
CS-Poincaré	δ_p^n	$D = 2n+1$	$ISO(D-1,1)$
CS-AdS	$\dfrac{l^{2p-D}}{(D-2p)} \dbinom{n}{p}$	$D = 2n+1$	$SO(D-1,2)$
CS-dS	$(-1)^p \dfrac{l^{2p-D}}{(D-2p)} \dbinom{n}{p}$	$D = 2n+1$	$SO(D,1)$

6.5   Finite Action and the Beauty of Gauge Invariance

Classical symmetries of a theory are defined as invariances modulo surface terms in the action because they are usually assumed to vanish in the variations. This is true for boundary conditions that keep the values of the fields fixed: Dirichlet conditions. In a gauge theory, however, it may be more relevant to fix gauge invariant properties at the boundary — like the curvature — and this would not be a Dirichlet boundary condition, but rather a condition of the Neumann type.

On the other hand, it is also desirable to have an action which has a finite value, when evaluated on a physically observable configuration e.g., on a classical solution. This is not just for the sake of elegance, it is a necessity in the study of semiclassical thermodynamic properties of the theory, which is particularly relevant for a theory possessing black holes with interesting thermodynamic features. Moreover, quasi-gauge invariant actions defined on an infinitely extended spacetime are potentially ill-defined. This is because under gauge transformations, the boundary terms generated might give infinite contributions to the action integral. This would not only cast doubt on the meaning of the action itself, but it would violently contradict the wish to have a gauge invariant action principle.

Changing the action by a boundary term may seem innocuous but it is a delicate business. The empirical fact is that adding a total derivative to a Lagrangian in general changes the expression for the conserved Noether charges, and again, possibly by an infinite amount. The conclusion from this discussion is that some regularization principle must be in place in order for the action to be finite on physically interesting configurations. That would ensure that the action remains finite under gauge transformations and yields well-defined conserved charges.

In Ref. [77] it is shown that the action has an extremum when the field equations hold, and is finite on classically interesting configurations if the AdS action (5.47) is supplemented with a boundary term of the form

$$B_{2n} = -\kappa n \int_0^1 dt \int_0^t ds\, \epsilon\theta e \left(\widetilde{R} + t^2\theta^2 + s^2 e^2 \right)^{n-1}, \qquad (6.3)$$

where $\widetilde{R}$ and θ are the intrinsic and extrinsic curvatures of the boundary, respectively. The resulting action attains an extremum for boundary conditions that fix the extrinsic curvature of the boundary. In that reference it is also shown that this action principle yields finite charges (mass, angular momentum) without resorting to ad-hoc regularizations or background subtractions. It can be asserted that in this case, as in many others, the demand of gauge invariance is sufficient to cure other seemingly unrelated problems.

The boundary term (6.3) that guarantees the convergence of the action and charges turns out to have other remarkable properties. It makes the action gauge invariant, and not just quasi-invariant, under gauge transformations that keep the curvature constant (AdS geometry at the boundary), and the form of the extrinsic curvature fixed at the boundary. The condition of having a fixed AdS asymptotic geometry is natural for localized matter distributions such as black holes. Fixing the extrinsic curvature, on the other hand, implies that the connection approaches a fixed reference connection at infinity in a prescribed manner.

On closer examination, this boundary term can be seen to convert the action into the integral of a *transgression form*. A transgression form is a gauge invariant object whose exterior derivative yields the difference of two characteristic classes $\mathcal{Q}_{2n}$ (A) [8],

$$dT_{2n-1}(A, \bar{A}) = \mathcal{Q}_{2n}(A) - \mathcal{Q}_{2n}(\bar{A}), \qquad (6.4)$$

where A and $\bar{A}$ are two connections in the same Lie algebra. There is an explicit expression for the transgression form in terms of the Chern-Simons forms for A and $\bar{A}$,

$$\mathcal{T}_{2n+1}(A, \bar{A}) = \mathcal{C}_{2n+1}(A) - \mathcal{C}_{2n+1}(\bar{A}) + d\mathfrak{B}_{2n}(A, \bar{A}). \tag{6.5}$$

The last term in the r.h.s. is uniquely determined by the condition that the transgression form be invariant under simultaneous gauge transformations of both connections throughout the entire manifold M

$$A \to A' = \Lambda^{-1}A\Lambda + \Lambda^{-1}d\Lambda, \tag{6.6}$$

$$\bar{A} \to \bar{A}' = \bar{\Lambda}^{-1}\bar{A}\bar{\Lambda} + \bar{\Lambda}^{-1}d\bar{\Lambda}, \tag{6.7}$$

with the matching condition at the boundary,

$$\bar{\Lambda}(x) = \Lambda(x), \quad \text{for} \quad x \in \partial M. \tag{6.8}$$

It can be seen that the boundary term in (6.3) is precisely the boundary term $\mathfrak{B}_{2n}$ in the transgression form. The interpretation now presents some subtleties. Clearly one is not inclined to duplicate the fields by introducing a second dynamically independent set of fields ($\bar{A}$), with exactly the same couplings, spin, gauge symmetry and quantum numbers as A.

One possible interpretation is to view the second connection as a non-dynamical reference field. This goes against the well-established principle that in the action every quantity that is not a coupling constant, mass parameter, or numerical coefficient like the dimension or a combinatorial factor, should correspond to a dynamical quantum variable [43]. Even if one accepts the existence of this uninvited guest, an explanation would be needed to justify why it is not seen in nature.

An alternative interpretation could be to assume that the spacetime is duplicated and we happen to live on one of the two parallel worlds where A is present, while $\bar{A}$ is in the other. These two worlds need to share the same boundary, for it would be very hard to justify the condition (6.8) otherwise. This picture gains support from the fact that the two connections do not interact on M, but only at the boundary, through the boundary term $\mathfrak{B}(A, \bar{A})$. An obvious drawback of this interpretation is that the action for $\bar{A}$ has the wrong sign, and therefore it will lead to ghosts or rather unphysical negative energy states.

Although this sounds like poor science fiction, it suggests a more reasonable option proposed in Ref. [91]: Since the two connections do not interact, they could have support on two completely different but cobordant manifolds. Then, the action really reads,

$$I[A, \bar{A}] = \int_M \mathcal{C}(A) + \int_{\bar{M}} \mathcal{C}(\bar{A}) + \int_{\partial M} \mathfrak{B}(A, \bar{A}), \tag{6.9}$$

where the orientations of M and $\bar{M}$ are appropriately chosen so that they join at their common boundary with $\partial M = -\partial \bar{M}$.

This interpretation shows a picture that we live in a region of spacetime (M) characterized by the dynamical field A. At the boundary of our region, ∂M, there exists another field with identical properties as A and matching gauge symmetry. This second field $\bar{A}$ extends on to a cobordant manifold $\bar{M}$, to which we have no direct access except through the interaction of $\bar{A}$ with our A. If the spacetime we live in is asymptotically AdS, this could be a reasonable scenario since the boundary is causally connected to the bulk and can be easily viewed as the common boundary of two (or more) asymptotically AdS spacetimes [92].

The boundary term (6.3) explicitly depends on the extrinsic curvature, which would be inadequate in a formulation that uses the intrinsic properties of the geometry as boundary data. The Dirichlet problem, in which the boundary metric is specified is an example of intrinsic framework. The standard regularization procedure in the Dirichlet approach uses counterterms that are covariant functions of the intrinsic boundary geometry. In Ref. [93], the transgression and counterterms procedures have been explicitly compared in asymptotically AdS spacetimes using an adapted Fefferman–Graham coordinate frame [94]. Both regularization techniques are shown to be equivalent in that frame, and they differ at most by a finite counterterm that does not change the Weyl anomaly. The regularization techniques were also extended for nonvanishing torsion, yielding a finite action principle and holographic anomalies in five dimensions in Ref. [95].

Chapter 7

Black Holes, Particles and Branes

By embedding the Lorentz group into one of its parents the de Sitter or anti-de Sitter group, or its more distant relative, the Poincaré group one can generate CS gravity theories for the spacetime geometry in each odd dimension. These gauge theories are based on the affine (ω) and metric (e) features of the spacetime manifold as the only dynamical fields of the system. The CS theories have no dimensionful couplings and are the natural generalization of gravity in $2+1$ dimensions. From here one can go on to study those theories, analyzing their classical solutions, their cosmologies and the black holes that inhabit them.

Another possible extension would be to investigate embeddings in other, larger groups. One could embed the Lorentz group $SO(D-1,1)$ in any $SO(n,m)$, if $n \geq D-1$ and $m \geq 1$, and contractions of them, analogous to the limit of vanishing cosmological constant that yields the Poincaré group. The results in this direction are rather dull. There are also some accidents like the (local) identity between $SO(3)$ and $SU(2)$, which occur occasionally, but those happy coincidences are rare.

The only other natural generalization of the Lorentz group into a larger group are direct products with other groups that yields theories constructed as mere sums of Chern-Simons actions, which is not very interesting. The reason for this boring scenario, as we shall see in the next chapters, is connected to the so-called *No-Go Theorems* [96]. Luckily, there is a very remarkable (and at the time revolutionary) way out of this murky situation, provided by *supersymmetry*. In fact, despite all the propaganda and false expectations generated by this unobserved symmetry, its most remarkable feature and possibly its only lasting effect in our culture is that it provides a natural way to unify the symmetries of spacetime and internal symmetries, like the gauge invariance of electrodynamics, the weak and the strong interactions. These issues will be treated in detail in the next chapter.

7.1   Chern-Simons Black Holes

As we have seen, the Lovelock action defined by (4.2) involves, apart from the cosmological and Newton's constants, $[(D-3)/2]$ additional arbitrary, dimensionful constants. This arbitrariness can give rise to dynamical indeterminacies with obscure physical interpretation, including unpredictable/indefinite time evolution and black holes with up to $[D/2]$ distinct horizons. This is the main reason to restrict ourselves to the Lovelock action whose underlying symmetry extends the Lorentz group. More precisely, we will consider the AdS CS gravity theories defined in odd dimensions where all the couplings are fixed in terms of Newton's constant and the cosmological constant.

7.1.1   *Three-dimensional black holes*

Before going to the general case $(D \geq 5)$, let us summarize the three-dimensional black hole.

In $2+1$ dimensions, the Einstein equations completely fix the Riemann tensor. This implies that the curvature at a given spacetime point is fully determined by the energy-momentum tensor $T^\mu{}_\nu$ at that point. In particular, the vacuum equations in the presence of a cosmological constant force the Riemann tensor to take the form

$$R^{\mu\nu}{}_{\alpha\beta} = c \left(\delta^\mu_\alpha \delta^\nu_\beta - \delta^\nu_\alpha \delta^\mu_\beta \right), \tag{7.1}$$

and therefore the geometry everywhere is either de Sitter $(c > 0)$, anti-de Sitter $(c < 0)$ or flat $(c = 0)$, where the cosmological constant c is a fixed parameter in the action. Therefore, there are no fluctuations of the geometry in vacuum space (no gravitational waves), in stark contrast with what happens in $3+1$ or higher dimensions. For instance, the four-dimensional Schwarzschild (-AdS) solution in vacuum has a curvature that is nowhere constant, approaching Minkowski (or AdS) spacetime for $r \to \infty$ and a curvature that diverges for $r \to 0$. In a vacuum region of $2+1$ spacetime, on the other hand, all solutions are locally the same.

The absence of propagating degrees of freedom in three-dimensional GR (no gravitons) was interpreted as the absence of gravitational attraction as in Newtonian gravity. Without Newtonian attraction, gravitational collapse and black holes were widely believed not to exist in three dimensional spacetime.

Spherically symmetric point sources where known in flat [97] and in constant curvature $(2+1)$-dimensional spacetime [98]. For positive cosmological constant a static spherically symmetric solution had been studied [99], although due to the "wrong" sign of Λ that solution has no horizon and therefore is not a black hole but a naked singularity. For an extended historical and technical discussion of gravitation in $2+1$ dimensions, see Ref. [100].

With all these negative indications, the existence of a $(2+1)$-dimensional black hole solution deserves an explanation. The most important fact to understand in connection with the black hole presented in Ref. [101] is that it is locally three-dimensional AdS spacetime, but differs from the global AdS space by an identification that changes the topology and therefore the black hole geometry is not just AdS in disguise [102]. The two geometries are not in the same topological class.

The black hole is produced by identifying points in AdS$_3$ connected by a Killing vector field, in an analogous operation to the one that produces a cylinder by identifying two parallel lines in a plane. This identification does not change the local (flat) geometry, but the resulting manifold cannot be continuously deformed to the original plane: these two manifolds belong to different topological classes. Another example of identification by a Killing vector is the one that matches two radial lines in a disc, producing a cone. In this second example, the Killing vector that produces the identification is a rotation by a fixed angle around the center, which creates a singularity at $r = 0$, the fixed point of the Killing vector field.

The Killing vector that produces the $2 + 1$ black hole is spacelike outside the horizon $(r > r_+)$, timelike inside the horizon $(r < r_+)$ and null on the horizon $(r = r_+)$. The central singularity is produced by the need to remove the region $r < 0$ that contains closed timelike curves. Hence, $r = 0$ is not a curvature singularity but simply the boundary beyond which a causally well-defined spacetime does not exist.

The $2 + 1$ black hole shares most of the features of its $3 + 1$ relative: it has a similar causal structure, a horizon hiding the singularity for positive mass M, and a naked singularity if $M < 0$. The general solution that can also have angular momentum (J) reads

$$ds^2 = -f^2 dt^2 + \frac{dr^2}{f^2} + r^2(N dt + d\phi)^2, \tag{7.2}$$

where $f^2 = -M + \frac{r^2}{l^2} + \frac{J^2}{4r^2}$, $N = -\frac{Jl}{2r^2}$, with $|J| \leq Ml$. The solution can also be extended to include electric and magnetic charge, although the electrically charged and spinning solution is much harder to obtain [103]. The electrically charged black hole is a solution for the field equations whose action includes the Maxwell term, so it is not strictly a purely gravitational system.

These black holes have well-defined temperature and entropy, and satisfy all the laws of black hole thermodynamics. Additionally, these black holes bend the trajectories of test particles around them [104] and can also be produced by gravitational collapse of two-dimensional dust clouds [105–107] and "stars" [108]. For a thorough review, see Ref. [100].

Finally, we note that one can add a magnetic charge to this black hole (7.2) by simply including a $U(1)$ connection

$$A^{(g)} = \frac{g}{2\pi} d\phi, \tag{7.3}$$

where g is the magnetic charge. The reason $A^{(g)}$ can be added without affecting the geometry of the solution is because it has vanishing field strength everywhere $(F^{(g)} = dA^{(g)} = 0$ for $r > 0)$, and does not contribute to the energy-momentum of the electromagnetic field. In fact, in any local patch that does not include the origin $A^{(g)}$ can be considered as a pure gauge artifact that could be gauged away $(A^{(g)} = d[g\phi/2\pi])$. The fact that the topology of the spatial section, $\mathbb{R}^2 - \{0\}$, is not simply connected means that $A^{(g)}$ is not just a gauge artifact and therefore the magnetic charge g is actually a physical (gauge invariant) attribute of the solution.

An important lesson can be extracted from the last observation. When the space-time geometry has a singularity that must be removed, this singular set can be the support of a current that is the source of a gauge field. In the example above, the current generated by the magnetic charge is the two form

$$J = g\delta(x, y)dx \wedge dy, \tag{7.4}$$

that sources the $U(1)$ connection through the coupling

$$\int J \wedge A. \tag{7.5}$$

This coupling can be seen to be gauge (quasi-) invariant because J is gauge invariant and closed, so that $\delta(J \wedge A) = J \wedge \delta A = J \wedge d\Omega = d(J \wedge \Omega)$. The expression (7.5) is just the ordinary minimal coupling $j^\mu A_\mu$, where j^μ is the dual current, $(1/2)\epsilon^{\mu\nu\rho}J_{\nu\rho}$. The dynamics of A that admits (7.3) as solution is given by the CS form AdA, so the magnetically charged black hole is an extremum of the action

$$\frac{\kappa}{l} \int_{M^3} \epsilon_{abc} \left(R^{ab} + \frac{1}{3l^2} e^a e^b \right) e^c + \int_{M^3} A \wedge dA - g \int_{\Gamma^1} A, \tag{7.6}$$

where Γ^1 is the one-dimensional line $r = 0$ that corresponds to the worldline of the magnetic charge.

The action (7.6) has the appeal of simplicity and beauty. It is the combination of three Chern-Simons forms: the three-dimensional CS forms for $SO(2,2)$ and $U(1)$ connections, and the one-dimensional CS form for the minimal coupling. Note that neither the $U(1)$ kinetic term nor the minimal coupling require the metric, unlike the case of the Maxwell action, which cannot even be written if the background geometry were not defined.

7.1.2   *Higher dimensional C-S black holes*

AdS-CS gravities in $D = 2n + 1$ dimensions, whose Lagrangians are given by (5.49), give rise to the following field equations

$$\left(R^{a_1 a_2} + l^{-2} e^{a_1} e^{a_2} \right) \cdots \left(R^{a_{2n-1} a_{2n}} + l^{-2} e^{a_{2n-1}} e^{a_{2n}} \right) \epsilon_{a_1 a_2 \cdots a_{2n+1}} = 0. \tag{7.7}$$

These equations also admit black hole solutions whose metric generalizes the static three-dimensional expression (7.2) [61]

$$ds^2 = -f^2(r)dt^2 + f^{-2}(r)dr^2 + r^2 d\Omega_{2n-1}^2, \tag{7.8}$$

with $f^2 = 1 - (M + 1)^{\frac{1}{n}} + (r/l)^2$. These solutions represent static, spherically symmetric configurations with a unique event horizon at the root of $f^2 = 0$, that is $r_+ = l\sqrt{(M + 1)^{\frac{1}{n}} - 1}$. The Riemann curvature is given by

$$R^{0r}_{0r} = -l^{-2}, \tag{7.9}$$

$$R^{0j}_{0k} = -l^{-2}\delta^j_k = R^{rj}_{rk}, \tag{7.10}$$

$$R^{jk}_{ml} = -l^{-2}\left[-l^{-2} + \frac{(M + 1)^{\frac{1}{n}}}{r^2} \right] \delta^{[jk]}_{[ml]}, \tag{7.11}$$

where the indices j, k, l, m label the angular coordinates in the $D - 2$ sphere. For $D > 3$ and $M \neq -1$ these geometries have a curvature singularity surrounded by the horizon. For $M = -1$ these geometries have constant negative curvature (AdS spacetime), and if $D = 3$ the components of (7.11) are absent and the metric reduces to the static BTZ solution (metric (7.2) with $J = 0$).

As discussed in Section 5.3, CS gravities are a particular subclass of the Lovelock series, obtained for a very particular choice of Lovelock coefficients a_p in odd dimensions. This choice has the peculiarity that the resulting theory acquires an enhanced gauge symmetry and all the dimensionful parameters of the theory can be absorbed in the fields, which are now all components of a dimensionless gauge connection. It means that the CS gravity theories are scale-invariant and the dimensionful constants such as the mass of a particular black hole solution is an integration constant that becomes dimensionful in the metric representation.

Static spherically symmetric solutions in Lovelock theories were studied long ago by several authors [109–112]. The question then naturally arises whether the solutions (7.8) exhibit some special feature that would not be shared by the black holes in a generic Lovelock theory. The first difference between black holes in CS gravity and a static, spherically symmetric solution in a generic Lovelock theory is the number of event horizons: one in the first case and up to $[(D - 1)/2]$ in the generic case. In fact, there exist infinitely many ways to select the a_p in such a way to produce a unique event horizon, but the CS choice is more special in another way.

In Ref. [64] a family of Lovelock theories with a unique effective cosmological constant possessing black holes with a single event horizon was studied. These theories are labeled by an integer k and, in odd dimensions, they reduce to CS choice for $D = 2k + 1$. All the black holes were asymptotically AdS spacetimes with finite mass $M(r_+)$, temperature $T(r_+)$, and well-defined thermodynamics, making the comparison much easier. For the CS black holes both $M(r_+)$ and $T(r_+)$ are monotonic functions and the specific heat $C = \partial M / \partial T$ is always positive. For the generic cases $(k \neq [D - 1]/2)$, $T(r_+) \to \infty$ for $r_+ \to 0$ and for $r_+ \to \infty$, with a single minimum at a critical value $r_+ = r_c$ (see Fig. 7.1). The specific heat is negative for $r_+ < r_c$, goes to $-\infty$ as r_+ approaches r_c^-, while it is positive for $r_+ > r_c$, goes to ∞ as r_+ approaches r_c^+. It means that for $k \neq [D - 1]/2$, even with a single horizon, generic

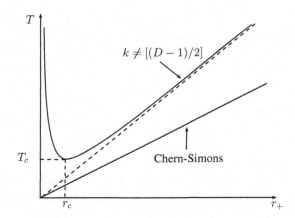

Fig. 7.1   Black holes in Lovelock theories with a unique cosmological constant and a single horizon labeled by an integer k. For $k = (D-1)/2$, which can only occur in odd dimensions, the action becomes that of the AdS-CS theory. The temperature as a function of the horizon radius changes dramatically between the generic and the CS cases.

black holes have negative specific heat for small r_+, which implies that small black holes get hotter as they emit radiation and loose mass, making them emit more and become increasingly hot, in an explosive process. For large r_+, instead, these black holes have positive specific heat and therefore reach thermal equilibrium with their surroundings. CS black holes, instead, have positive specific heat for any r_+ and any M and therefore they always attain thermal equilibrium with the surrounding temperature (see Fig. 7.2). In that way, CS black holes behave like ideal gases that could be used as perfect thermometers.

7.2   Naked Singularities, Particles and Branes

If the mass parameter of the black hole solution is taken to negative values, the horizon disappears and the singularity at the origin becomes causally connected to observers everywhere, it is a *naked singularity*. Naked singularities have a bad reputation and might be scary [113], but this particular type is quite a harmless singularity, a conical defect.

7.2.1   *The darker side of the 3D black holes*

The metric (7.2) has constant curvature and remains so in the limit $r \to 0$ for all values of J and $M \neq 0$. For $M \geq 0$ (BTZ black hole), the singularity at $r = 0$ is a boundary as already discussed. For $-1 < M < 0$, the Lorentz connection is singular at the origin because a vector parallel-transported along a closed curve gets rotated by an angle $(2n\pi\Delta)$ where n is the winding number of the curve around $r = 0$, and $0 < \Delta < 1$ is a fixed coefficient related to the mass. In other words, the singularity

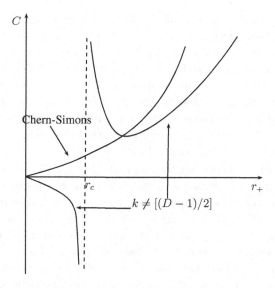

Fig. 7.2 The specific heat of the Lovelock $k \neq [(D-1)/2]$ and CS ($k = (D-1)/2$) cases respectively. The generic case is explosive for $r_+ \to 0$, while the CS case behaves like an ideal gas, always reaching equilibrium with a heat bath.

is of the same nature as the one found at the tip of a cone. This *conical singularity* is clearly a curvature singularity but of a mild type, it essentially indicates that $r = 0$, like the tip of a cone, does not belong to the spacetime manifold [114].

In order to see this, consider the solution (7.2) with $M < 0$ and $J = 0$ (the case $J \neq 0$ will be discussed next). By an appropriate rescaling of the coordinates, the metric is that of AdS$_3$ with a central conical singularity,

$$ds^2 = -[r^2/l^2 + 1]dt^2 + [r^2/l^2 + 1]^{-1}dr^2 + (1 - \Delta)^2 r^2 d\phi^2, \qquad (7.12)$$

where $-\infty < t < \infty$, $0 \leq r < \infty$, $0 \leq \phi \leq 2\pi \sim 0$ and $0 < \Delta < 1$. The proper length of the circle of small radius ϵ around $r = 0$ is $2\pi\epsilon(1 - \Delta)$. Since for small radius $r \sim \epsilon$ is the proper radial distance, the deficit angle is identified as $2\pi\Delta$. This angular defect at the origin that is revealed in the connection,

$$\omega^a{}_b: \quad \omega^0{}_1 = rdt, \quad \omega^0{}_2 = 0, \quad \omega^1{}_2 = [1 - \Delta](r^2/l^2 + 1)^{1/2}d\phi. \qquad (7.13)$$

The coordinate ϕ is not defined at the origin. Hence $d\phi$ is not a well-defined differential form in an open set containing the origin and therefore $\omega^1{}_2$ is ill-defined at $r = 0$. This can also be seen in the curvature two-form,

$$R^{ab} = -\frac{1}{l^2}e^a e^b + \pi\Delta \left(\delta^a_1\delta^b_2 - \delta^a_2\delta^b_1\right)\delta(x^1)dx^1 \wedge \delta(x^2)dx^2, \qquad (7.14)$$

which has been computed using the formula valid for a cone, $dd\phi = a\delta(x^1)dx^1 \wedge \delta(x^2)dx^2$, where a is the angular deficit at the tip [115], that can also be computed by parallel-transporting a vector along a closed loop.

Equation (7.14) describes a locally AdS spacetime with a naked curvature singularity at the origin. The singularity at $r = 0$ is the position of a point particle, represented by the δ-distribution in (7.14). Alternatively, a singularity like the apex of a cone, is a point where the manifold fails to be differentiable, there is no tangent space and therefore is not properly a point of the manifold; this point is removed from every spatial section of constant t. In this light, this naked singularity defines an excluded set from the manifold, quite different from the naked singularity obtained in the $3 + 1$ Schwarzschild solution with negative mass parameter, where the curvature diverges in the open set $\{0 < r < \epsilon\}$.

Conical singularities were identified as point particles long before the discovery of the black hole in $2 + 1$ dimensions [97, 98]. From the geometric point of view, conical singularities and black holes have the same AdS asymptotics and the same global isometries — time translations $\xi_t = \partial_t$ and axial rotations $\xi_\phi = \partial_{\phi^-}$, and the same local geometry. Both types of objects are solutions of the same field equations, endowed with the same set of conserved charges — mass M and angular momentum $J-$, but occupying different sections of the spectrum corresponding to different topologies. Black holes and point particles are extrema of the same gravitational action principle. Black holes and point particles are made of the same stuff, they just have different topologies.

For zero angular momentum, the deficit angle and the mass are related [114] by $M = -(1 - \Delta)^2$, with a range $-1 < M < 0$, where $M = -1$ corresponds to AdS ($\Delta = 0$, no angular deficit) and $M = 0$ is the zero mass black hole ($\Delta = 1$, angular deficit of 2π). Point particles have an angular deficit corresponding to their masses and when they merge their angular deficits add up. If their added angular deficits exceeds the 2π threshold, a black hole is formed [116]. For masses below the AdS vacuum, $M < -1$, the solution represents a negative angular deficit that is, an *angular excess*. These angular excesses are also δ-like curvature singularities and since the angular deficits for these singularities are additive, the angular excesses should be regarded as antiparticles, with AdS space as the vacuum. In the flat case, an angular excess produces a surface of "Elizabethan geometry" similar to the ruff collar in fashion around 1600 [117]. Unlike angular deficits, angular excesses can be arbitrarily large.

Point particles, antiparticles and black holes are distinct states of the same gravitation theory in which the different topologies result from the different choices for the integration constants in the solutions. There is no need to assume a different action principle to include these "sources" unless the evolution of the point particle is governed by some other interaction, as discussed next.

7.2.2 *Spinning point particles*

So far, we have discussed a static *zero-brane*, a particle at rest at $r = 0$. It is also possible to conceive a spinning particle as a solution in the same way that the

Kerr black hole generalizes the static solution. As shown in Ref. [103], the spinning configuration can be obtained by boosting the static one, $(t, r, \phi) \rightarrow (t', r', \phi')$, given by

$$t' = \frac{t - \omega l \phi}{\sqrt{1 - \omega^2}}, \quad \phi' = \frac{\phi - \omega t/l}{\sqrt{1 - \omega^2}}, \quad r'^2 = r^2 - \frac{M\omega^2}{1 - \omega^2}, \tag{7.15}$$

where $M = -(1 - \Delta)^2$, and $\omega = J/(1 - \Delta)^2 l$ takes values in the range $-1 < \omega < 1$. This boosting takes a solution into another solution because it is compatible with the isometries that are preserved by the identification in AdS that produces the black hole [102].

It works as follows: the $2 + 1$ black hole is obtained by identifying points in AdS space that are connected by one of the six global Killing vectors in AdS. This leaves the local geometry intact, while the global isometry group is reduced to those isometries that commute with the Killing vector. The isometry group of AdS$_3$ is $SO(2, 2)$, which has rank 2; then, identifying by a Killing vector ∂_u leaves as commuting isometries ∂_u and one more Killing vector, say, ∂_v. These correspond to the generators of the Cartan subalgebra of $SO(2, 2)$. In the $2 + 1$ black hole these two commuting generators are ∂_t and ∂_ϕ. Therefore, any linear combination of them is also an isometry and in particular, a rotation in the (t, ϕ)-plane, with a convenient redefinition $r \rightarrow r'(r)-$ such as (7.15) also defines an admissible identification. The result in the case of the black hole is (7.2), while for the point particle it is given by the same formula but with $M = -(1 - \Delta)^2$.

An interesting aspect of the boost (7.15) is that the new mass and angular momentum in the spinning case are related to the original (static) mass as

$$M' = \frac{1 + \omega^2}{1 - \omega^2} M \quad J' = \frac{2\omega l}{1 - \omega^2} M. \tag{7.16}$$

Since $\omega^2 \leq 1$, it implies that $M'^2 - J'^2/l^2 = M^2 \geq 0$. In other words, the combination $M^2 - J^2/l^2 \geq 0$ remains invariant under (7.15) and these boosts connect states lying of the same hyperbola $M^2 - J^2/l^2 = M_0^2 \geq 0$. The states with $M^2 - J^2/l^2 \leq 0$, on the other hand, are not reachable by these transformations and correspond to the analogue of tachyonic particle states. In the same vein, the extremal black holes and particle states $M^2 - J^2/l^2 = 0$, correspond to light-like particles obtained by infinite boosts, $\omega \rightarrow \pm 1$.

7.2.3 *Charged point particles*

A point particle as described above can also be electromagnetically charged. This charge couples naturally to a $U(1)$ connection A defined on the 1-dimensional worldline,

$$I_{EM} = q \int_\Gamma A_\mu dx^\mu. \tag{7.17}$$

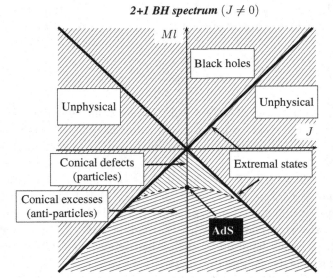

2+1 BH spectrum $(J \neq 0)$

Fig. 7.3   The $2 + 1$ black hole $(Ml \geq |J|)$ and point particle $(Ml \leq -|J|)$ spectra. The lines $Ml = \pm|J|$ represent extremal black holes and extremal point particles. The spinning states can be reached by boosting the static configurations but keeping the quantity $M^2l - J^2$ fixed. Hence the regions $|J| \geq |M|l$ cannot be obtained in this way and are analogous to tachyonic states.

The integration over the worldline Γ can also be expressed as a volume integral,

$$I_{EM} = \int_{AdS_3} \mathcal{J} \wedge A, \tag{7.18}$$

where $\mathcal{J}$ is the current two-form defined by the conical singularity, dual to the electric current,

$$\mathcal{J} = q\delta(x^1)dx^1 \wedge \delta(x^2)dx^2. \tag{7.19}$$

Expression (7.18) is the minimal coupling between a point charge and the electromagnetic field, $\int j^\mu(x)A_\mu dV$, where j^μ is the electromagnetic current density. The usual form of the conservation law, $\partial_\mu j^\mu = 0$, is replaced by $d\mathcal{J} = 0$ and invariance of this coupling under a gauge transformation $A \to A + d\Omega$, is guaranteed in both cases by current conservation.

The minimal coupling (7.17) is remarkable in its simplicity: it is gauge invariant in spite of the fact that it depends explicitly on the connection A and not on the gauge-invariant field strength $F = dA$. Moreover, it is metric-independent, which makes it Lorentz-invariant in Minkowski space or invariant under general coordinate transformations in a curved background.

Strictly speaking, (7.17) is not invariant but quasi-invariant under gauge transformations, it changes by a boundary term $q[A(+\infty) - A(-\infty)]$. These properties of the minimal coupling are not surprising since it is the integral of a CS form: if $\mathcal{C}_{2n+1} = A(dA)^n$ is the Abelian $(2n+1)$-CS form, then $\mathcal{C}_1 = A$ is the corresponding $(0+1)$-CS form.

The charge q can be electric or magnetic, depending on the action that defines the dynamics of A. For instance, if the system is three-dimensional, the electromagnetic Lagrangian can be taken as the CS form, AdA or the Maxwell form, $*F \wedge F$. In the first case, q is a source of the magnetic field (monopole), while in the second it is an electric charge. In the first case the electromagnetic field satisfies the equation $F = \mathcal{J}$ and the solution is $A = (q/2\pi)d\phi$, is the vector potential produced by a magnetic pole. Note that since the CS form does not involve the metric, this solution does not couple to the spacetime geometry and is the same if the background Minkowski, AdS$_3$, a $2 + 1$ black hole, or any of the point particles described by conical singularities.

7.2.4 $(D-3)$-*branes as conical defects in* D *dimensions*

The 0-brane previously discussed is an *orbifold*, i.e., a quotient space M/K, where M is a smooth manifold and K is a Killing vector field with a fixed point, $K = \Delta\partial_\phi$. This is equivalent to removing an angular wedge of magnitude Δ from the (x^1, x^2)-plane, so that (t, r, ϕ) is identified with $(t, r, \phi + 2\pi - \Delta)$. The vector K is spacelike everywhere in AdS$_3$ but is not defined on the set of fixed points of K, $\Gamma = \{x_0 = (t, 0, \phi)\}$.

This singular set Γ is precisely the support of the Dirac δ distribution in the curvature two-form. At the singularity the tangent manifold is not defined; the conical singularity is a singularity of the tangent bundle, but other structures, like the metric for instance, are not affected by this. The differentiable manifold, fully equipped with the tangent structure, is the complement of the singular set, AdS$_3 - \Gamma$. Excising the singular set does not modify the local geometry in the rest of the manifold and therefore the identification only changes the spacetime topology by excluding from the manifold the set occupied by the worldline of the point particle.

A naked singularity produced by identification in a manifold of constant curvature does not produce a region where the curvature grows infinitely, as in the central curvature singularity in four-dimensional Schwarzschild black hole. It means that conical or causal singularities generated in this form could not emit unbounded amounts of energy or become an endless source of paradoxes.

A similar identification by a rotation with a fixed angle in a given plane can be performed in all spacetimes of dimensions $D > 2$. This produces a $(D-2)$-dimensional singular set, also referred to as a codimension 2 set. This conical defect in D dimensions is also located at the fixed point of the invariant plane of the Killing vector. This singularity can be regarded as the worldvolume of $(D-3)$ brane. For $D = 2n+1$, this is a $(2n-2)$-brane that can couple to a $(2n-1)$-CS form $\mathcal{C}_{2n-1}(A)$. The case discussed above is particular: a 0-brane coupled to the CS one-form A in (7.18) [118–122]. The $(2n-1)$-CS form that couples to the worldhistory of the brane can correspond to any particular Lie algebra and in that case the brane is charged with respect to the corresponding gauge group.

This identification again changes the topology produced by the removal of the singular set that represents the worldvolume of the brane, but does not change the local geometry. For $D > 3$, the naked singularity produced in this way does not relate to the D-dimensional black hole by occupying the negative range of its mass spectrum as is the case for the three-dimensional conical singularity. For $D \neq 3$ the black holes are not locally constant curvature geometries and therefore black holes and conical defects of codimension 2 cannot share the same local geometry. Hence, for $D > 3$ it is not true that a large enough conical defect produces a black hole.

7.3   Chern-Simons Branes

The preceding discussion shows that a $2p$-brane evolving embedded in an ambient spacetime of any dimension $D > 2p + 1$ can also be the support of a CS form in complete analogy to the way an electric charge couples to the electromagnetic potential. These *CS-branes* become charged sources for the connection field in the ambient space.

7.3.1   *The simplest CS branes*

The simplest case is the point electric charge already discussed, and the next example is a $U(1)$-charged 2-brane Σ, an ordinary static 2-dimensional surface. Let us further assume Σ to be static, orientable, simply connected, smooth and without boundary. This membrane traces a $2+1$ dimensional history in spacetime, the tube $\Gamma = \Sigma \times \mathbb{R}$, which supports an abelian CS form,

$$I_{Brane} = \theta \int_\Gamma AdA, \tag{7.20}$$

where θ is the electric charge of the membrane, and $\Gamma \subset M$ is an isometric embedding. This brane is a source of an electromagnetic field Minkowski space, which for illustrative purposes will be described by Maxwell's equations.

Since Σ has no boundary, it is typically the boundary of some compact region $\Sigma = \partial B_0$, $B_0 \subset \mathbb{R}^3$. By the same token, Γ is the boundary of $M_0 \equiv B_0 \times \mathbb{R}$. Then, applying stokes theorem, I_{Brane} can be written as

$$I_{\text{Brane}} = \theta \int_{M_0} F \wedge F, \tag{7.21}$$

which is readily recognized as the integral of the Pontryagin topological density, commonly referred to as the θ-term [123]. This term does not affect Maxwell's equations either inside or outside the tube Γ, because in both regions I_{Brane} is just a surface term.

The above expression is presumably important in quantum field theory as it gives different relative weights to states in different topological sectors of the theory. Here (7.21) can be regarded as having the region R_0 filled with some material with identical dielectric and magnetic properties as empty space, but whose quantum vacuum is characterized by a parameter $\theta \neq 0$. Since the two regions are separated by a physical membrane, the vacuum inside need not be equivalent to the one outside.

It turns out that such materials do exist and are commonly known as *topological insulators* [85]. As shown in Ref. [124] the CS term (7.21) adds an extra contribution to Maxwell's equations at the interface,

$$\nabla \cdot \mathbf{E} = \theta \delta(\Sigma) \mathbf{B} \cdot \mathbf{n}, \qquad (7.22)$$

$$\nabla \times \mathbf{B} - \partial_t \mathbf{E} = \theta \delta(\Sigma) \mathbf{E} \times \mathbf{n}. \qquad (7.23)$$

The effect of the CS term is to induce a "charge density" proportional to $\mathbf{B} \cdot \mathbf{n}$ at the surface, and a "current density" proportional to $\mathbf{E} \times \mathbf{n}$. It produces a rotation in the polarization plane of the reflected and transmitted waves at the interface.

7.3.2 *General CS branes*

As emphasized in previous chapters, one of the characteristic features of a CS form is that all its coefficients are fixed dimensionless rational numbers determined uniquely by the Lie (super-)algebra and the dimension of spacetime. If any of these coefficients is changed, gauge invariance is lost and the result is not very interesting. It means that the theory does not admit a natural splitting into a "free" part and a "small" perturbative correction, and there is no coupling parameter that can be set equal to zero in order to define the unperturbed starting point. In other words, a perturbative expansion in powers of a coupling constant around some classical configuration of the "unperturbed" theory is not defined.

The absence of dimensionful adjustable coefficients may be a blessing in the sense that the theory cannot receive corrections under renormalization. On the other hand, it is not clear how to probe the response of the system via interactions. The observation that CS systems can be supported on membranes makes it conceivable that membranes of different ranks embedded in the same ambient spacetime could interact through their gauge fields that propagate in the embedding space. This is a generalization of the interaction between point charges mediated by the electromagnetic field. In fact, branes can interact through the CS forms supported in them, as illustrated by the following examples.

2p Brane Embedded in an Odd-Dimensional Spacetime

This possibility is an extension of the conical defect discussed in §7.2. A specially interesting possibility occurs if the ambient spacetime itself is governed by a CS theory, in which case both the coupling to the sources and their interactions are described by CS forms. In $D = 2n + 1$ dimensions, a $2p + 1$ CS form couplings supported by the worldhistories of the $2p$-branes with $0 < p < n$ have been considered in [120–122]. These higher dimensional branes provide consistent couplings between CS gravity (or CS supergravities) and localized sources. Since these couplings are manifestly gauge invariant, these branes do not break local Lorentz or AdS symmetry (or the corresponding supersymmetries) of the D-dimensional theory.

Brane Embedded in an Even-Dimensional Spacetime

Examples of this sort are the topological insulators discussed in Section 7.3. If two membranes of this sort are present, they interact through the gauge field in the ambient space, whose dynamics could be described by a Maxwell, Yang-Mills, Einstein-Hilbert- or more generally, Lovelock- Lagrangian.

An extremely simple situation of this kind occurs if the brane is a topological defect embedded in a space whose "action" is purely topological. This happens, for instance, if the Lagrangian of the even-dimensional ambient spacetime is the Euler or Pontryagin form. Suppose an identification is made by means of a rotational Killing vector or magnitude θ on a certain plane. It was shown in Ref. [125] that the excision of the angular sector produces a topological defect and the induced action at the defect is dynamical. In this way, a four-dimensional gravitation can be considered a defect in a six-dimensional topological spacetime.

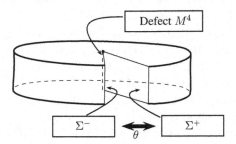

Fig. 7.4 Example of an identification in a six-dimensional spacetime by a rotational Killing vector of magnitude θ in a 2-dimensional plane. Identifying the 5-dimensional surfaces Σ^+ and Σ^- gives rise to a 4-dimensional topological defect M^4. If the Lagrangian for the 6-dimensional ambient space is the Euler form, the induced dynamics in M^4 is given by the Einstein-Hilbert action.

Chapter 8

Supersymmetry and Supergravity

We have dealt so far with the possible ways in which pure gravity can be extended by relaxing three standard assumptions of General Relativity:

  i) Notion of parallelism derived from metricity,
  ii) Four-dimensional spacetime, and
  iii) Einstein-Hilbert Lagrangian $\sqrt{-g}[R - 2\Lambda]$ only.

Instead, we demanded:

  i′) Metricity and parallelism described by dynamically independent fields, e^a and $\omega^a{}_b$,
  ii′) D-dimensional spacetime, and
  iii′) D-form Lagrangian function of e^a, $\omega^a{}_b$ and their exterior derivatives; quasi-invariant under local Lorentz rotations in tangent space.

In this way, a family of Lagrangians containing higher powers of the curvature and torsion multiplied by arbitrary dimensionful coefficients is obtained. *In odd dimensions only*, the embarrassing presence of arbitrary constants can be cured by enlarging the symmetry into the de Sitter, anti-de Sitter or Poincaré groups. This makes the theory gauge-invariant under the enhanced symmetry and simultaneously fixes all parameters in the Lagrangian. The result is a highly nonlinear Chern-Simons theory of gravity, invariant under local $SO(D,1)$, $SO(D-1,2)$, or $ISO(D-1,1)$ transformations in the tangent space.

We now turn to the problem of enlarging the contents of the theory by allowing for bosonic and fermionic fields as well as transformations that mix them, *supersymmetry*. This will have two effects: it will incorporate fermions and fermionic generators into the picture, and will further enlarge the spacetime symmetry by including additional bosonic generators. These new bosonic symmetries are required by consistency –the closure of the algebra–, a most important consequence of supersymmetry.

8.1   Supersymmetry

Supersymmetry (**SUSY**) is a curious symmetry: most theoreticians are willing to accept it as a legitimate feature of nature, although it has never been experimentally observed in spite of systematic and intensive searches over the past four decades. The reason for its popularity rests on its compelling unifying beauty that makes one feel it would be a pity if such an elegant feature were not realized in nature somehow.

The most interesting aspect of SUSY is its ability to combine *spacetime symmetries* — like Lorentz or Poincaré invariances — with other *internal gauge* symmetries like the $SU(3) \times SU(2) \times U(1)$ invariance of the standard model. Thus SUSY supports the hope that it could be possible to establish a logical connection between spacetime and internal invariances. SUSY makes more natural the idea that these different bosonic symmetries might be related and somehow need each other. In this way it could explain why some internal symmetries are observed and others are not. The most important lesson from supersymmetry may turn out not to be the unification of bosons and fermions, but the extension of the bosonic symmetry.

Supersymmetry transforms **bosons** (integer spin particles) into **fermions** (half integer spin particles), and vice-versa. These two types of particles obey very different statistics and satisfy completely different dynamical equations, so the existence of a symmetry that connects them would be rather surprising.[1]

The simplest supersymmetric theories combine bosons and fermions on equal footing, rotating them into each other under SUSY transformations in a way analogous to the mixing of the components of a vector under rotations. This is possibly the most intriguing, and uncomfortable aspect of supersymmetry: the blatant fact that bosons and fermions play such radically different roles in nature and exhibit such different features means that SUSY is not manifest around us, and therefore, it must be strongly broken at the scale of our observations. Unbroken SUSY would imply the existence of fermionic carriers of interactions and bosonic constituents of matter as partners. None of these two types of particles have been observed. Moreover, according to the simplest form of SUSY, the fermions and bosons related by the symmetry must have the same masses and charges, only differing in spin by half a unit. It is remarkable that a symmetry for which there is no evidence at all be postulated and excuse its absence arguing that it must be broken at an

---

[1] Bosons obey Bose-Einstein statistics and, like ordinary classical particles, there is no limit to the number of them that can occupy the same state. Fermions, on the other hand, cannot occupy the same quantum state more than one at a time, which gives rise to the Fermi-Dirac statistics. All elementary particles are either bosons or fermions and they play different roles: fermions like electrons, protons, neutrons, leptons and quarks are the constituents of matter, while the four known fundamental interactions of nature are described by gauge fields resulting from the exchange of bosons, the photon, the gluon, the $W^{\pm}$ and Z^0, and the graviton.

energy beyond present experimental range. Several mechanisms for SUSY breaking have been postulated but there is no consensus at present as to which is the correct one [126].

In spite of its failure to describe the observed reality, SUSY gained acceptance in the high energy community mainly because it offers the possibility of taming the ultraviolet divergences of many field theories. It was observed early on that the UV divergences of the bosons were often canceled out by divergences coming from the fermionic sector. This possibility seemed particularly attractive in the case of a quantum theory of gravity, and in fact, it was shown that in a supersymmetric extension of general relativity, dubbed supergravity (**SUGRA**) the ultraviolet divergences at the one-loop level exactly canceled (see Ref. [127] and references therein). This is another remarkable feature of SUSY: local (gauge) SUSY is not only compatible with gravity, consistency of local SUSY *requires* gravity.

From an algebraic point of view, SUSY is the simplest nontrivial way to enlarge the Poincaré group, unifying spacetime and internal symmetries, thus circumventing an important obstruction found by S. Coleman and J. Mandula [96]. The obstruction, also called *no-go theorem*, roughly states that if a physical system has a symmetry described by a Lie group **G**, which contains the Poincaré group and some other internal symmetry, then the corresponding Lie algebra must be a direct sum of the form $\mathcal{G} = \mathcal{P} \oplus \mathcal{S}$, where $\mathcal{P}$ and $\mathcal{S}$ are the Poincaré and the internal symmetry algebras, respectively [128, 129]. Supersymmetry is nontrivial because the algebra *is not* a direct sum of the spacetime and internal symmetries. The way the no-go theorem is circumvented is to have both commutators (antisymmetric product, $[\cdot, \cdot]$) and anticommutators (symmetric product, $\{\cdot, \cdot\}$), forming what is known as a *graded Lie algebra*, also called super Lie algebra, or simply superalgebra [129, 130].

8.1.1 *Superalgebras*

A superalgebra has two types of generators: bosonic, $\mathbf{B}_i$, and fermionic, $\mathbf{F}_\alpha$. They are closed under the (anti-)commutator operation, which follows the general pattern

$$[\mathbf{B}_i, \mathbf{B}_j] = C_{ij}^k \mathbf{B}_k, \tag{8.1}$$

$$[\mathbf{B}_i, \mathbf{F}_\alpha] = C_{i\alpha}^\beta \mathbf{F}_\beta, \tag{8.2}$$

$$\{\mathbf{F}_\alpha, \mathbf{F}_\beta\} = C_{\alpha\beta}^i \mathbf{B}_i. \tag{8.3}$$

The generators of the spacetime symmetry group are included in the bosonic sector, and the $\mathbf{F}_\alpha$'s are the supersymmetry generators. Relations (8.1) indicate that the bosonic generators form a subalgebra, which is not expected to be simple or semi-simple because it should contain the product of spacetime and internal symmetries that commute with each other. The second set of commutators states that the fermionic sector should be in some vector-like irreducible representation of the bosonic symmetry. This is typical of fermionic fields, since they are in a vector-like (spinor) representation of the Lorentz group and form a vector multiplet of the gauge group (charged under $U(1)$, colored under $SU(3)$, etc.). The third relation is the one that ties together the different parts, making the entire family a nontrivial unity where all the pieces are needed to close the superalgebra.

Operators satisfying relations of the form (8.1)–(8.3), are still required to satisfy a consistency condition, the super-Jacobi identity, which is required by associativity,

$$[\mathbf{G}_I, [\mathbf{G}_J, \mathbf{G}_K]_\pm]_\pm + (-)^{\sigma(JKI)} [\mathbf{G}_J, [\mathbf{G}_K, \mathbf{G}_I]_\pm]_\pm$$
$$+ (-)^{\sigma(KIJ)} [\mathbf{G}_K, [\mathbf{G}_I, \mathbf{G}_J]_\pm]_\pm = 0. \qquad (8.4)$$

Here $\mathbf{G}_I$ represents any generator in the algebra, $[\mathbf{A}, \mathbf{B}]_\pm = \mathbf{AB} \pm \mathbf{BA}$, where the sign is chosen according the bosonic or fermionic nature of the operators in the bracket, and $\sigma(JKI)$ is the number of permutations of fermionic generators necessary for $(IJK) \to (JKI)$.

This algebra, however, does not close for an arbitrary bosonic group. Given a Lie group with a set of bosonic generators, it is not always possible to find a set of fermionic generators to enlarge the algebra into a closed superalgebra unless a number of extra bosonic generators are brought in. It is also often the case that apart from extra bosonic generators, a collection of $\mathcal{N}$ fermionic operators are needed to close the algebra. This usually works for some values of $\mathcal{N}$ only, but in some cases there is simply no supersymmetric extension at all [131]. This happens, for example, if one starts with the de Sitter group, which has no supersymmetric extension in general [129]. For this reason, we will restrict to AdS theories in the following sections. The general problem of classifying all possible superalgebras that extend the classical Lie algebras has been discussed in Ref. [132].

8.1.2 *Supergravity*

The name supergravity (**SUGRA**) applies to any of a number of supersymmetric theories that include gravity in their bosonic sectors.[2] The invention/discovery of supergravity in the mid 70's came about with the spectacular announcement that some ultraviolet divergent graphs in pure gravity were canceled by the inclusion of their supersymmetric counterparts. For some time it was hoped that the supersymmetric extension of GR could be renormalizable. However, it was eventually realized that SUGRAs too might turn out to be nonrenormalizable [133].

---

[2]Some authors would reserve the word *supergravity* for supersymmetric theories whose gravitational sector is the Einstein-Hilbert (**EH**) Lagrangian. This narrow definition seems untenable for dimensions $D > 4$ in view of the variety of possible gravity theories beyond EH. Our point of view here is that there can be more than one system that can be called supergravity, although the relation between these supergravities and the standard theory is not direct.

Standard SUGRAs are not gauge theories for a group or a supergroup, and the local (super-)symmetry algebra closes naturally on shell only, that is, if the field equations hold. In some cases the algebra could be made to close off shell by force, at the cost of introducing some nondynamical *auxiliary fields*. These extra fields are not guaranteed to exist for all D and $\mathcal{N}$ [134], and still the theory would not have a fiber bundle structure since the base manifold is identified with part of the fiber. Whether it is the lack of fiber bundle structure the ultimate reason for the nonrenormalizability of gravity remains to be proven. It is certainly true, nevertheless, that if GR could be formulated as a gauge theory, the chances for its renormalizability would clearly improve. At any rate, now most high energy physicists view supergravity as an effective theory obtained from string theory in some limit. In string theory, eleven dimensional supergravity is seen as an effective theory obtained from ten dimensional string theory at strong coupling [135]. In this sense supergravity would not be a fundamental theory and therefore there is no reason to expect it to be renormalizable.

8.2   From Rigid Supersymmetry to Supergravity

Rigid (also called global) SUSY is a symmetry in which the group parameters are constant throughout spacetime. In particle physics spacetime is usually assumed to have fixed Minkowski geometry and the relevant SUSY is the supersymmetric extension of the Poincaré algebra in which the supercharges are "square roots" of the generators of spacetime translations, $\{\bar{\mathbf{Q}}, \mathbf{Q}\} \sim \Gamma \cdot \mathbf{P}$. The extension of this to a local symmetry can be done by substituting the momentum $\mathbf{P}_\mu = i\partial_\mu$ by the generators of spacetime diffeomorphisms, $\mathcal{H}_\mu$, and relating them to the supercharges by $\{\bar{\mathbf{Q}}, \mathbf{Q}\} \sim \Gamma \cdot \mathcal{H}$. The resulting theory has a local supersymmetry algebra which only closes on-shell [127]. As we discussed above, the problem with on-shell symmetries is that they are unlikely to survive quantization.

An alternative approach for constructing the SUSY extension of the spacetime symmetry is to work on the tangent space rather than on the base spacetime manifold. This point of view is natural if one recalls that spinors are defined relative to a local frame on the tangent space and not as tensors of the coordinate basis. In fact, spinors provide an irreducible representation for $SO(N)$, but not for $GL(N)$, which describes infinitesimal general coordinate transformations. The basic strategy is to reproduce the $2+1$ "miracle" in higher dimensions. This idea was applied in five dimensions [75], as well as in higher odd dimensions [3–5].

8.2.1   *Standard supergravity*

In its simplest version, supergravity was conceived in the early 70s, as a quantum field theory whose action included the Einstein-Hilbert term, representing a massless spin-2 graviton, plus a Rarita-Schwinger term describing a massless spin-3/2 particle, the gravitino [136]. These fields would transform into each other under local supersymmetry. Later on, the model was refined to include more "realistic" features, like matter couplings, enlarged symmetries, higher dimensions with their corresponding reductions to 4D, cosmological constant, etc., [127]. In spite of the number of variations on the theme, a few features remained as the hallmark of SUGRA, which were a reflection of this history. In time, these properties have become a sort of identikit of SUGRA, although they should not be taken as a set of necessary postulates. Among these features, three of them will be relaxed in our construction:

  (i) Gravity described by the EH action (with or without cosmological constant),
 (ii) The spin connection and the vielbein are related through the torsion equation, and,
(iii) The fermionic and bosonic fields in the Lagrangian come in combinations such that they have equal number of propagating degrees of freedom.

The last feature is inherited from rigid supersymmetry in Minkowski space, where particles form vector representations of the Poincaré group labeled by their spin and mass, and the matter fields form vector representations of the internal groups (multiplets). This is justified in a Minkowski background where particle states are represented by in- and out- plane waves in a weakly interacting theory. This argument, however, breaks down if the Poincaré group is not the spacetime symmetry, as it happens in asymptotically AdS spacetimes and other cases like $1+1$ dimensions with broken translational invariance [137].

The argument for the matching between fermionic and bosonic degrees of freedom goes as follows: The generator of translations in Minkowski space, $P_\mu = (E, \mathbf{p})$, commutes with all symmetry generators, therefore an internal symmetry should only mix particles of equal mass. Since supersymmetry changes the spin by $1/2$, a supersymmetric multiplet must contain, for each bosonic eigenstate of the Hamiltonian $|E>_B$, a fermionic one with the same energy, $|E>_F = \mathbf{Q} |E>_B$, and vice versa. Thus, it seems natural that in supergravity this would still be the case. In AdS space, however, the momentum operator is not an abelian generator and does not necessarily commute with the supersymmetry generator $\mathbf{Q}$. Another limitation of this assumption is that it does not consider the possibility that the fields belong to a different representation of the Poincaré or AdS group, such as the adjoint representation.

Also implicit in the argument for counting the degrees of freedom is the usual assumption that the kinetic terms are those of a minimally coupled gauge theory, a condition that does not hold in CS theories. Apart from the difference in background, which requires a careful treatment of the unitary irreducible representations of the asymptotic symmetries [138], the counting of degrees of freedom in CS theories follows a completely different pattern [17] from the counting for the same connection 1-forms in a YM theory [50].

The other two issues concern the purely gravitational sector and are dictated by economy, but in view of Lovelock's theorem, there is no reason to adopt (i), and the fact that the vielbein and the spin connection are dynamically independent fields on equal footing makes assumption (ii) unnatural. Furthermore, the spin connection dependent on the vielbein introduces the inverse vielbein in the action and thereby entangling the action of the spacetime symmetries defined on the tangent space. The fact that the supergravity generators do not form a closed off-shell algebra may be traced back to these assumptions.

8.2.2 *AdS superalgebras*

In order to construct a supergravity theory that contains gravity with cosmological constant, a mathematically oriented physicist would look for the smallest superalgebra that contains the generators of the AdS algebra. This question was asked and answered, many years ago, at least for some dimensions $D = 2, 3, 4$ mod 8, in [131]. But having identified the superalgebra is not sufficient, we would also like to have a set of fields that carry an irreducible representation of the algebra and an action that realizes the symmetry.

Several supergravities are known for all dimensions $D \leq 11$ [139]. For $D = 4$, a supergravity action that includes a cosmological constant was first discussed in Ref. [140], however, finding a supergravity with cosmological constant in an arbitrary dimension is a nontrivial task. For example, the standard supergravity in eleven dimensions has been known for a long time [141], however, it is impossible to add a cosmological constant to it [142,143].[3] Moreover, although it was known to the authors of Ref. [141] that the supergroup containing the eleven dimensional AdS group is $SO(32|1)$, a gravity action which exhibits this symmetry was found almost twenty years later [4].

As previously discussed, given an algebra, a connection can be defined including one connection component for each generator in the algebra. In odd dimensions this is sufficient to construct a quasi-invariant CS form. This strategy was extended in Ref. [4] to superalgebras and we will now discuss how this construction and its multiple generalizations can be carried out.

First, we present the construction of the superalgebras that contain the AdS algebra, $SO(D - 1, 2)$. We follow the idea of [131] but extending the method to the cases $D = 5, 7$ and 9 as well [4]. The crucial observation is that the Dirac matrices provide a natural representation of the AdS algebra in any dimension. Thus, the

---

[3]Strictly speaking, what is shown in Refs. [142,143] is the impossibility of making a perturbative deformation of the theory to accommodate a cosmological constant. The supergravities discussed below are not perturbatively connected to the standard form of supergravity.

AdS connection $\mathbf{W}$ can be written in this representation as $\mathbf{W} = e^a \mathbf{J}_a + \frac{1}{2}\omega^{ab}\mathbf{J}_{ab}$, where

$$\mathbf{J}_a = \frac{1}{2}(\Gamma_a)^\alpha_\beta, \tag{8.5}$$

$$\mathbf{J}_{ab} = \frac{1}{2}(\Gamma_{ab})^\alpha_\beta. \tag{8.6}$$

Here Γ_a, $a = 1, \ldots, D$ are $m \times m$ Dirac matrices, where $m = 2^{[D/2]}$ (here $[s]$ is the integer part of s), and $\Gamma_{ab} = \frac{1}{2}[\Gamma_a, \Gamma_b]$. These two classes of matrices form a closed commutator subalgebra (the AdS algebra) of the **Dirac algebra**. The Dirac algebra is obtained by taking all the antisymmetrized products of Γ matrices

$$\mathbf{I}, \Gamma_a, \Gamma_{a_1 a_2}, \ldots, \Gamma_{a_1 a_2 \cdots a_D}, \tag{8.7}$$

where

$$\Gamma_{a_1 a_2 \cdots a_k} = \frac{1}{k!}(\Gamma_{a_1}\Gamma_{a_2} \cdots \Gamma_{a_k} \pm [permutations]).$$

For even D, the matrices in the set (8.7) are all linearly independent, but for odd D they are not, because $\Gamma_{12\cdots D} = \sigma \mathbf{I}$ and therefore half of them are proportional to the other half. Hence, the dimension of this algebra, i.e., the number of independent matrices of the form (8.7), is $m^2 = 2^{2[D/2]}$. This representation provides an elegant way to generate all $m \times m$ matrices (note that $m = 2^{[D/2]}$ is not *any* number).

8.3 Fermionic Generators

The supersymmetric extension of a given Lie algebra is a mathematical problem whose solution lies in the general classification of superalgebras [132]. Instead of approaching this problem as a general question of classification of irreducible representations, we will take a more practical course, by identifying the representation we are interested in from the start. This representation is the one in which the bosonic generators take the form (8.5) and (8.6). The simplest extension of the algebra generated by those matrices is obtained by the addition of one row and one column, as

$$\mathbf{J}_a = \begin{bmatrix} \frac{1}{2}(\Gamma_a)^\alpha_\beta & 0 \\ 0 & 0 \end{bmatrix}, \tag{8.8}$$

$$\mathbf{J}_{ab} = \begin{bmatrix} \frac{1}{2}(\Gamma_{ab})^\alpha_\beta & 0 \\ 0 & 0 \end{bmatrix}. \tag{8.9}$$

The generators associated to the new entries would have only one spinor index. Let us call $\mathbf{Q}_\gamma$ ($\gamma = 1, \ldots, m$) the generator that has only one nonvanishing entry in the γ-th row of the last column,

$$\mathbf{Q}_\gamma = \begin{bmatrix} 0 & \delta_\gamma^\alpha \\ -C_{\gamma\beta} & 0 \end{bmatrix} \qquad \alpha, \beta = 1, \ldots, m. \tag{8.10}$$

Since this generator carries a spinorial index (γ), it is in a spin-1/2 representation of the Lorentz group.

The entries of the bottom row ($C_{\gamma\beta}$) will be so chosen as to produce the smallest supersymmetric extensions of AdS. There are essentially two ways to restrict the dimension of the representation compatible with Lorentz invariance: *chirality* and *reality*. In odd dimensions there is no chirality because the corresponding "γ_5" is proportional to the identity matrix. Reality instead can be defined in any dimension and refers to whether a spinor and its conjugate are proportional up to a constant matrix, $\bar{\psi} = \mathbf{C}\psi$, or more explicitly,

$$\bar{\psi}^\alpha = C^{\alpha\beta}\psi_\beta. \tag{8.11}$$

A spinor that satisfies this condition is said to be Majorana, and $\mathbf{C} = (C^{\alpha\beta})$ is called the charge conjugation matrix. This matrix is assumed to be invertible, $C_{\alpha\beta}C^{\beta\gamma} = \delta_\alpha^\gamma$, and plays the role of a metric in the space of Majorana spinors.

Using the form (8.10) for the supersymmetry generator, its Majorana conjugate $\bar{\mathbf{Q}}$ is found to be

$$\bar{\mathbf{Q}}^\gamma = C^{\alpha\beta}\mathbf{Q}_\beta$$

$$= \begin{bmatrix} 0 & C^{\alpha\gamma} \\ -\delta_\beta^\gamma & 0 \end{bmatrix}. \tag{8.12}$$

The matrix $\mathbf{C}$ can be viewed as performing a change of basis $\psi \to \psi^{\mathrm{T}} = \mathbf{C}\psi$, which in turn corresponds to the change $\Gamma \to \Gamma^{\mathrm{T}}$. Now, since the Clifford algebra for the Dirac matrices,

$$\{\mathbf{\Gamma}^a, \mathbf{\Gamma}^b\} = 2\eta^{ab}, \tag{8.13}$$

is also obeyed by their transpose, $(\mathbf{\Gamma}^a)^{\mathrm{T}}$, these two algebras must be related by a change of basis, up to a sign,

$$(\mathbf{\Gamma}^a)^{\mathrm{T}} = \eta\mathbf{C}\mathbf{\Gamma}^a\mathbf{C}^{-1} \quad \text{with } \eta^2 = 1. \tag{8.14}$$

The basis of the Clifford algebra (8.13) for which an operator $\mathbf{C}$ satisfying (8.14) exists, is called *the Majorana representation*. This last equation is the defining relation for the charge conjugation matrix, and whenever it exists, it can be chosen to have definite parity,[4]

$$\mathbf{C}^{\mathrm{T}} = \lambda\mathbf{C}, \quad \text{with} \quad \lambda = \pm 1. \tag{8.15}$$

---

[4]The Majorana reality condition can be satisfied in any D provided the spacetime signature is such that, if there are s spacelike and t timelike dimensions, then $s - t = 0, 1, 2, 6, 7$ mod 8 [129, 130]. Thus, for Lorentzian signature, Majorana spinors can be defined unambiguously only for $D = 2, 3, 4, 8, 9$, mod 8.

8.3.1   *Closing the algebra*

We already encountered the bosonic generators responsible for the AdS transformations (8.5) and (8.6), which have the general form required by (8.1). It is straightforward to check that commutators of the form $[\mathbf{J},\mathbf{Q}]$ turn out to be proportional to $\mathbf{Q}$, in agreement with the general form (8.2). What is by no means trivial is the closure of the anticommutator $\{\mathbf{Q},\mathbf{Q}\}$ as in (8.3). Direct computation yields

$$\{\mathbf{Q}_\gamma,\mathbf{Q}_\lambda\}^\alpha_\beta = \begin{bmatrix} 0 & \delta^\alpha_\gamma \\ -C_{\gamma\rho} & 0 \end{bmatrix}\begin{bmatrix} 0 & \delta^\rho_\lambda \\ -C_{\lambda\beta} & 0 \end{bmatrix} + (\gamma \leftrightarrow \lambda)$$

$$= -\begin{bmatrix} \delta^\alpha_\gamma C_{\lambda\beta} + \delta^\alpha_\lambda C_{\gamma\beta} & 0 \\ 0 & C_{\gamma\lambda} + C_{\lambda\gamma} \end{bmatrix}. \tag{8.16}$$

The form of the lower diagonal piece immediately tells us that unless $C_{\gamma\lambda}$ is antisymmetric, the right hand side of (8.16) cannot be a linear combination of $\mathbf{J_a}$, $\mathbf{J_{ab}}$ and $\mathbf{Q}$. In that case, new bosonic generators with nonzero entries in this diagonal block will be required to close the algebra (and possibly more than one). This relation also shows that the upper diagonal block is a collection of matrices $\mathbf{M}_{\gamma\lambda}$ whose components take the form

$$(M_{\gamma\lambda})^\alpha_\beta = -(\delta^\alpha_\gamma C_{\lambda\beta} + \delta^\alpha_\lambda C_{\gamma\beta}).$$

Multiplying both sides of this relation by C (lowering the index α), one finds

$$(CM_{\gamma\lambda})_{\alpha\beta} = -(C_{\alpha\gamma}C_{\lambda\beta} + C_{\alpha\lambda}C_{\gamma\beta}), \tag{8.17}$$

which is symmetric in $(\alpha\beta)$. This means that the bosonic generators can only include those matrices in the Dirac algebra such that, when multiplied by $\mathbf{C}$ on the left ($\mathbf{CI}$, $\mathbf{C\Gamma}_a$, $\mathbf{C\Gamma}_{a_1a_2}$, ..., $\mathbf{C\Gamma}_{a_1a_2\cdots a_D}$) turn out to be symmetric. The other consequence of this is that, if one wants to have the AdS algebra as part of the superalgebra, both $\mathbf{C\Gamma}_a$ and $\mathbf{C\Gamma}_{ab}$ should be symmetric matrices. Now, multiplying (8.14) by $\mathbf{C}$ from the right, we have

$$(\mathbf{C\Gamma}_a)^{\mathrm{T}} = \lambda\eta\mathbf{C\Gamma}_a, \tag{8.18}$$

which means that we need

$$\lambda\eta = 1. \tag{8.19}$$

From (8.14) and (8.15), it can be seen that

$$(\mathbf{C\Gamma}_{ab})^{\mathrm{T}} = -\lambda\mathbf{C\Gamma}_{ab},$$

which in turn requires

$$\lambda = -1 = \eta.$$

It means that $\mathbf{C}$ is antisymmetric ($\lambda = -1$) and then the lower diagonal block in (8.16) vanishes identically. However, the values of λ and η cannot be freely chosen but are fixed by the spacetime dimension as shown in the following Table 8.1,

Table 8.1

D	λ	η
3	-1	-1
5	-1	$+1$
7	$+1$	-1
9	$+1$	$+1$
11	-1	-1

This table shows that the simple cases occur for dimensions 3 mod 8, while for the remaining cases life is harder. For $D = 7$ mod 8 the need to match the lower diagonal block with some generators can be satisfied quite naturally by including several spinors labelled with a new index, ψ_i^α, $i = 1, \ldots, \mathcal{N}$, and the generator of supersymmetry should also carry the same index. It means that there are actually $\mathcal{N}$ supercharges or, as it is usually said, the theory has an extended supersymmetry ($\mathcal{N} \geq 2$). For $D = 5$ mod 4 instead, the superalgebra can be made to close in spite of the fact that $\eta = +1$ if one allows for complex spinor representations, which is a particular form of extended supersymmetry since now $\mathbf{Q}_\gamma$ and $\overline{\mathbf{Q}}^\gamma$ become independent generators.

So far we have only given some restrictions necessary to close the algebra so that the AdS generators appear in the anticommutator of two supercharges. As we have observed, in general, apart from $\mathbf{J}_a$ and $\mathbf{J}_{ab}$, other matrices will occur in the r.h.s. of the anticommutator of $\mathbf{Q}$ and $\overline{\mathbf{Q}}$ which extends the AdS algebra into a larger bosonic algebra. This happens even in the cases in which the supersymmetry is not extended ($\mathcal{N}=1$).

The bottom line of this construction is that the supersymmetric extension of the AdS algebra for each odd dimension falls into one of these families:

- $D = 3$ mod 8 (Majorana representation, $\mathcal{N} \geq 1$),
- $D = 7$ mod 8 (Majorana representation, even $\mathcal{N}$), and
- $D = 5$ mod 4 (complex representations, $\mathcal{N} \geq 1$ [$2\mathcal{N}$ real spinors]).

The corresponding superalgebras[5] were computed by van Holten and Van Proeyen for $D = 2, 3, 4$ mod 8 in Ref. [131], and in the other cases, in Refs. [4,5]:

Table 8.2

D	S-Algebra	Conjugation Matrix
3 mod 8	$osp(m\|\mathcal{N})$	$C^T = -C$
7 mod 8	$osp(\mathcal{N}\|m)$	$C^T = C$
5 mod 4	$usp(m\|\mathcal{N})$	$C^\dagger = C$

---

[5]These are the orthosymplectic $osp(p|q)$ (resp. unitary-symplectic $usp(p|q)$) Lie algebras. The corresponding Lie groups are defined as those that leave invariant a quadratic form $G_{AB}z^A z^B = g_{ab}x^a x^b + \gamma_{\alpha\beta}\theta^\alpha \theta^\beta$, where g_{ab} is a p-dimensional symmetric (resp. Hermitean) matrix and $\gamma_{\alpha\beta}$ is a q-dimensional antisymmetric (resp. anti-Hermitean) matrix.

Chapter 9

Chern-Simons Supergravities

In this chapter, we will be concerned with the CS supergravity actions for the supersymmetric extensions of the AdS or Poincaré algebras. As seen in the previous chapters, the local Lorentz symmetry of Lovelock gravity actions can be extended to a local Poincaré, dS or AdS symmetry by a judicious choice of the Lovelock coefficients. The resulting theories share interesting features since they are genuine CS gauge theories for the Poincaré, dS or AdS groups. Here, we will show how it is possible to construct the supersymmetric extensions of the CS AdS or Poincaré gravity actions in any odd dimension. The resulting supergravities actions differ from those of the standard supergravity on two main aspects. First is the fact that the fundamental field is played by a connection one-form and second, the supersymmetry transformations close off-shell without the need of introducing auxiliary fields.

As already mentioned, the generic construction for a CS Lagrangian in $D = 2n+1$ dimensions for a given Lie (super)algebra with generators $\boldsymbol{G}_A$ can be done in two steps. First, define a Lie-algebra-valued connection one-form with components $\boldsymbol{A} = A^A \boldsymbol{G}_A$ whose corresponding curvature two-form $\boldsymbol{F} = d\boldsymbol{A} + \boldsymbol{A} \wedge \boldsymbol{A} = F^A \boldsymbol{G}_A$ is also obviously expanded along the generators of the Lie algebra. Second, find an invariant tensor of rank $(n+1)$ denoted by $g_{A_1 \cdots A_{n+1}}$ and satisfying $\nabla g_{A_1 \cdots A_{n+1}} = 0$. If an explicit matrix representation is available, the trace $\langle \boldsymbol{G}_{A_1} \boldsymbol{G}_{A_2} \cdots \boldsymbol{G}_{A_p} \rangle$ provides an invariant tensor of rank p in the center of the algebra.[1] However, there is in general no guarantee that this $g_{A_1 \cdots A_p}$ does not vanish, and for some algebras there can be more than one invariant tensor of a given rank. Equipped with these two ingredients, a CS Lagrangian L_{CS} can be generically defined such that its exterior derivative is given by the contraction of the invariant tensor with the components of the field strength two-form as

$$dL_{CS} = g_{A_1 \cdots A_{n+1}} F^{A_1} \wedge \cdots \wedge F^{A_{n+1}}. \tag{9.1}$$

---

[1]In the case of a superalgebra, the invariant tensor is a *supertrace* [129].

As we have seen, the corresponding CS action is automatically quasi-invariant with respect to the gauge transformations defined by $\delta_\lambda A = d\lambda + [A, \lambda]$, where λ is the zero-form transformation parameter. Since we are interested in constructing CS supergravity actions, the main difficulty lies in finding the appropriate Lie super algebras and the invariant tensors that yield relevant physical actions.

For any semisimple algebra like AdS or its supersymmetric extensions, the trace (or supertrace) easily provides an invariant tensor $g_{A_1 \cdots A_{n+1}}$ and defines a CS Lagrangian. Therefore, it is enough to find a supersymmetric extension of the AdS algebra to construct an action quasi-invariant under local super-AdS transformations. On the other hand, for non-semisimple algebras like Poincaré or their supersymmetric extensions, the situation is much more difficult because there does not exist a systematic way to construct the invariant tensor. Finding a physically relevant invariant tensor is rendered difficult by two main reasons. In practice there are many possible super extensions and one has to "guess" the appropriate extension of the Poincaré algebra that yields a physically interesting theory. In addition, there is no systematic way to construct the invariant tensor, and finding it by solving the equation $\nabla g_{A_1 \cdots A_{n+1}} = 0$ is far from obvious.

It is for this reason that we opt for a constructive approach to the supersymmetric extension of the Poincaré CS action. The method is similar to the Noether procedure and its main lines can be summarized as follows. We start from the Poincaré CS action which is a gauge theory for the Poincaré algebra. We then extend the Poincaré algebra to its smallest extension by only adding the fermionic generators yielding the commonly known super Poincaré algebra. The supersymmetric gauge transformations for the vielbein, the spin connection and the gravitino will be read off from this superalgebra. The variation of the Poincaré CS action with respect to these supersymmetric transformations will require the introduction of a Rarita-Schwinger term in order to cancel part of this variation. As shown below, in three dimensions, the variation of the Poincaré CS action together with the Rarita-Schwinger term yield a total derivative, and hence a supersymmetric extension of the Poincaré action is possible for the super Poincaré algebra. On the other hand, for odd dimensions $D \geq 5$, in order for this procedure to close, the super Poincaré algebra must be extended with the introduction of an extra fifth-rank bosonic generator, and such algebra is usually dubbed as super five-brane algebra. Finally, with the explicit form of the CS supergravity action, we will write down the relevant invariant tensor.

9.1   AdS CS Supergravity Actions

For the construction of the AdS CS supergravity actions in odd dimensions, the situation is rendered simple because the AdS algebra and its supersymmetric extensions are semisimple algebras. Indeed, in this case, a possible invariant tensor can be given by the supertrace in the representation of the supersymmetric AdS algebra.

Here, we present the actions in three, five and eleven dimensions because, as shown in Table 5, these supersymmetries are quite different from each other and at the same time, each of these is the first of a family with similar features, (3 mod 8, 5 mod 4 and 7 mod 8). The origin of this peculiarity lies in a feature of the Clifford algebras, related to *Bott periodicity* of homotopy groups $\pi_k(O) = \pi_{k+8}(O)$, $\pi_k(Sp) = \pi_{k+8}(Sp)$, etc. [144].

9.1.1 *AdS$_3$ supergravity as a CS action in $D = 3$*

The peculiarity of the three-dimensional AdS case lies in the fact that there exist two locally AdS invariant Lagrangians, and both can be promoted to a CS supergravity action. The first AdS invariant Lagrangian preserves parity and is just the standard Einstein-Hilbert action with a cosmological constant. The other is the so-called *exotic AdS Lagrangian*, which involves torsion explicitly and is parity odd although it gives rise to parity-even field equations. In fact, as we have shown before, the field equations of both theories are equivalent. Here, we will present the supersymmetric extensions of the two AdS invariant actions for the smallest supersymmetric extension of the AdS group, namely the orthosymplectic group OSP(2|1) whose generators satisfy the following (anti)commutation relations

$$[J_a, J_b] = J_{ab}, \ [J_{ab}, J_c] = J_a \eta_{bc} - J_b \eta_{ac}, \ [J_{ab}, J_{cd}] = -J_{ac}\eta_{bd} + \cdots$$

$$[J_{[p]}, Q_\alpha] = \frac{1}{2}(\Gamma_{[p]})_\alpha^\beta Q_\beta, \ \{Q_\alpha, Q_\beta\} = 2(C\Gamma^a)_{\alpha\beta}J_a - 2(C\Gamma^{ab})_{\alpha\beta}J_{ab},$$

$$(9.2)$$

where $J_{[p]}$ denotes both AdS generators J_{ab} and J_a.

The connection 1−form expanded along the generators of the OSP(2|1) is

$$A = e^a J_a + \frac{1}{2}\omega^{ab}J_{ab} + \psi^\alpha Q_\alpha.$$

The supersymmetric transformations are gauge transformation with parameter $\lambda = \epsilon^\alpha Q_\alpha$ where ϵ^α is a Majorana spinor 0−form,

$$\delta e^a = 2\bar\epsilon\Gamma^a\psi, \qquad \delta\omega^{ab} = -2\bar\epsilon\Gamma^{ab}\psi, \qquad \delta\psi = \nabla\epsilon, \qquad (9.3)$$

where the spin covariant derivative is defined by

$$\nabla\psi := \left(d + \frac{1}{4}\omega_{ab}\Gamma^{ab} + \frac{1}{2}e_a\Gamma^a\right)\psi = D(\omega)\psi + \frac{1}{2}e_a\Gamma^a\psi. \qquad (9.4)$$

The supersymmetric extension of the three-dimensional Einstein-Hilbert action with negative cosmological constant is then given by

$$I_{G3} = \epsilon_{abc}\left(R^{ab} + \frac{1}{3}e^a e^b\right)e^c - 4\bar\psi\nabla\psi, \qquad (9.5)$$

and defines a CS gauge theory for the OSP(2|1) group.

On the other hand, the supersymmetric extension of the three-dimensional AdS invariant exotic action, involving torsion and invariant under the supersymmetric transformations (9.3), is given by

$$I_{T3} = L_3^*(\omega) + 2e_a T^a + 8\bar{\psi}\nabla\psi. \tag{9.6}$$

Here L_3^* is the Lorentz three-form satisfying $dL_3^*(\omega) = R^{ab}R_{ba}$ and T^a is the torsion two-form (3.37).

9.1.2 *AdS$_5$ supergravity as a CS action in D = 5*

The five-dimensional CS supergravity is obtained from the simplest supersymmetric extension of the AdS group in five dimensions. The gauge group is the unitary supergroup $SU(2,2|\mathcal{N})$, whose bosonic sector is given by

$$SU(2,2) \otimes SU(\mathcal{N}) \otimes U(1). \tag{9.7}$$

It might seem strange that there is no evidence of the presence of the Lorentz symmetry, but as we will see shortly, this part of the local gauge symmetry is hidden in the $SU(2,2)$ through its isomorphism with the AdS$_5$ group, $SO(4,2)$. The generators of the $su(2,2|\mathcal{N})$ algebra are spanned by

$$su(2,2): \qquad \boldsymbol{J}_{AB} = (\boldsymbol{J}_{ab}, \boldsymbol{J}_a)\,, \qquad (A, B = 0, \ldots, 5),$$

$$su(\mathcal{N}): \qquad \boldsymbol{T}_\Lambda\,, \qquad\qquad (\Lambda = 1, \ldots, \mathcal{N}^2 - 1),$$

$$\text{SUSY generators}: \quad \boldsymbol{Q}_r, \, \bar{\boldsymbol{Q}}^r\,, \qquad (r = 1, \ldots, \mathcal{N}),$$

$$u(1) - \text{central charge}: \boldsymbol{Z}.$$

A representation of the super algebra can be given by the following $(4+\mathcal{N}) \times (4+\mathcal{N})$ super matrices

$$\boldsymbol{J}_{AB} = \begin{pmatrix} \frac{1}{2}\,(\Gamma_{AB})^\beta_\alpha & 0 \\ 0 & 0 \end{pmatrix}, \quad \boldsymbol{T}_\Lambda = \begin{pmatrix} 0 & 0 \\ 0 & (\tau_\Lambda)^s_r \end{pmatrix},$$

$$\boldsymbol{Q}^\alpha_s = \begin{pmatrix} 0 & 0 \\ -\delta^r_s \delta^\alpha_\beta & 0 \end{pmatrix}, \quad \bar{\boldsymbol{Q}}^s_\alpha = \begin{pmatrix} 0 & \delta^s_r \delta^\beta_\alpha \\ 0 & 0 \end{pmatrix}, \tag{9.8}$$

$$\boldsymbol{Z} = \begin{pmatrix} \frac{i}{4}\delta^\beta_\alpha & 0 \\ 0 & \frac{i}{\mathcal{N}}\delta^s_r \end{pmatrix},$$

where $\alpha, \beta = 1, \cdots, 4; r, s = 1, \cdots, \mathcal{N}$ and the 4×4 matrices Γ_{AB} are defined in term of the Dirac matrices Γ_a as

$$\Gamma_{AB} = \left(\frac{1}{2}[\Gamma_a, \Gamma_b], \ \Gamma_{a5} = \Gamma_a \right).$$

Note that these Dirac matrices also provide a representation of the AdS group in 5 dimensions, which displays the isomorphism $SU(2,2) \simeq SO(4,2)$ mentioned above. Apart from the $su(2,2)$ and $su(\mathcal{N})$ subalgebras, generated by $\boldsymbol{J}_{AB}$ and τ_Λ respectively, the remaining non-vanishing (anti)commutation relations read

$$\left[\boldsymbol{J}_{AB}, \boldsymbol{Q}_s^\alpha \right] = -\frac{1}{2} \left(\Gamma_{AB} \right)_\beta^\alpha \boldsymbol{Q}_s^\beta , \qquad \left[\boldsymbol{T}_\Lambda, \boldsymbol{Q}_s^\alpha \right] = \left(\tau_\Lambda \right)_s^r \boldsymbol{Q}_r^\alpha ,$$

$$\left[\boldsymbol{J}_{AB}, \bar{\boldsymbol{Q}}_\alpha^s \right] = \frac{1}{2} \bar{\boldsymbol{Q}}_\beta^s \left(\Gamma_{AB} \right)_\alpha^\beta , \qquad \left[\boldsymbol{T}_\Lambda, \bar{\boldsymbol{Q}}_\alpha^s \right] = -\bar{\boldsymbol{Q}}_\alpha^r \left(\tau_\Lambda \right)_r^s ,$$

$$\left[\boldsymbol{Z}, \boldsymbol{Q}_s^\alpha \right] = -i \left(\frac{1}{4} - \frac{1}{\mathcal{N}} \right) \boldsymbol{Q}_s^\alpha , \quad \left[\boldsymbol{Z}, \bar{\boldsymbol{Q}}_\alpha^s \right] = i \left(\frac{1}{4} - \frac{1}{\mathcal{N}} \right) \bar{\boldsymbol{Q}}_\alpha^s ,$$

together with

$$\left\{ \boldsymbol{Q}_s^\alpha, \bar{\boldsymbol{Q}}_\beta^r \right\} = \frac{1}{4} \delta_s^r \left(\Gamma^{AB} \right)_\beta^\alpha \boldsymbol{J}_{AB} - \delta_\beta^\alpha \left(\tau^\Lambda \right)_s^r \boldsymbol{T}_\Lambda + i \delta_\beta^\alpha \delta_s^r \boldsymbol{Z} .$$

It is clear that $\mathcal{N} = 4$ is a special case in which the generator $\boldsymbol{Z}$ commutes with the fermionic generators and is therefore truly a central charge. As shown in Ref. [145], for $\mathcal{N}=4$ the algebra becomes the superalgebra $su(2,2|4)$ with a $u(1)$ central extension, or $su(2,2|4) \times u(1)$.

This superalgebra defines an invariant third rank tensor $g_{KLM} = Str[G_K G_L G_M]$, given by the supertrace in the explicit representation (9.8). This tensor is completely symmetric (resp. antisymmetric) in the bosonic (resp. fermionic) indices whose non-vanishing components read

$$g_{[AB][CD][EF]} = -\frac{1}{2} \varepsilon_{ABCDEF} , \qquad g_{z[AB][CD]} = -\frac{1}{4} \eta_{[AB][CD]} ,$$

$$g_{\Lambda_1 \Lambda_2 \Lambda_3} = -\gamma_{\Lambda_1 \Lambda_2 \Lambda_3} , \qquad g_{z\Lambda_1 \Lambda_2} = -\frac{1}{\mathcal{N}} \gamma_{\Lambda_1 \Lambda_2} ,$$

$$g_{[AB]\binom{\alpha}{r}\binom{s}{\beta}} = -\frac{i}{4} \left(\Gamma_{AB} \right)_\beta^\alpha \delta_r^s , \qquad g_{z\binom{\alpha}{r}\binom{s}{\beta}} = \frac{1}{2} \left(\frac{1}{4} + \frac{1}{\mathcal{N}} \right) \delta_\beta^\alpha \delta_r^s , \qquad (9.9)$$

$$g_{\Lambda\binom{\alpha}{r}\binom{s}{\beta}} = -\frac{i}{2} \delta_\beta^\alpha \left(\tau_\Lambda \right)_r^s , \qquad g_{zzz} = \frac{1}{\mathcal{N}^2} - \frac{1}{4^2} ,$$

where $\eta_{[AB][CD]} \equiv \eta_{AC} \eta_{BD} - \eta_{AD} \eta_{BC}$ and $\gamma_{\bar{K}\bar{L}}$ are the Killing metrics of $SU(2,2) \simeq SO(4,2)$ and $SU(\mathcal{N})$, respectively.

In order to construct the CS supergravity action for this supergroup, it is convenient to write the connection one-form along the generators of the superalgebra as

$$\boldsymbol{A} = e^a \boldsymbol{J}_a + \frac{1}{2} \omega^{ab} \boldsymbol{J}_{ab} + A^K \boldsymbol{T}_K + \left(\bar{\psi}^r \boldsymbol{Q}_r - \bar{\boldsymbol{Q}}^r \psi_r \right) + b \boldsymbol{Z} , \qquad (9.10)$$

The Chern-Simons Lagrangian for this gauge algebra is defined by the relation $dL = ig_{KLM} \boldsymbol{F}^K \boldsymbol{F}^L \boldsymbol{F}^M$, where $\boldsymbol{F} = d\boldsymbol{A} + \boldsymbol{A}^2$ is the (anti-hermitian) curvature. Using the explicit expression of the invariant tensor, one obtains the Lagrangian originally discussed in Ref. [75],

$$L = L_G(\omega^{ab}, e^a) + L_{su(\mathcal{N})}(A_s^r) + L_{u(1)}(\omega^{ab}, e^a, b) + L_F(\omega^{ab}, e^a, A_s^r, b, \psi_r), \quad (9.11)$$

with

$$L_G = \frac{1}{8}\epsilon_{abcde}\left[R^{ab}R^{cd}e^e + \frac{2}{3}R^{ab}e^ce^de^e + \frac{1}{5}e^ae^be^ce^de^e\right],$$

$$L_{su(\mathcal{N})} = -Tr\left[\boldsymbol{A}(d\boldsymbol{A})^2 + \frac{3}{2}\boldsymbol{A}^3 d\boldsymbol{A} + \frac{3}{5}\boldsymbol{A}^5\right],$$

$$L_{u(1)} = \left(\frac{1}{16} - \frac{1}{\mathcal{N}^2}\right)b(db)^2 + \frac{3}{4}\left[T^aT_a - R^{ab}e_ae_b - R^{ab}R_{ab}/2\right]b$$
$$+ \frac{3}{\mathcal{N}}F_s^r F_r^s b,$$

$$L_F = \frac{3}{2i}\left[\bar{\psi}^r\mathcal{R}\nabla\psi_r + \bar{\psi}^s\mathcal{F}_s^r\nabla\psi_r\right] + c.c. \quad (9.12)$$

where $A_s^r \equiv A^K(\boldsymbol{T}_K)_s^r$ is the $su(\mathcal{N})$ connection, F_s^r is its curvature, and the bosonic blocks of the supercurvature: $\mathcal{R} = \frac{1}{2}T^a\boldsymbol{\Gamma}_a + \frac{1}{4}(R^{ab} + e^ae^b)\boldsymbol{\Gamma}_{ab} + \frac{i}{4}db\boldsymbol{Z} - \frac{1}{2}\psi_s\bar{\psi}^s$, $\mathcal{F}_s^r = F_s^r + \frac{i}{\mathcal{N}}db\delta_s^r - \frac{1}{2}\bar{\psi}^r\psi_s$. The cosmological constant is $-l^{-2}$, and the $SU(2,2|\mathcal{N})$ covariant derivative ∇ acting on ψ_r is

$$\nabla\psi_r = D\psi_r + \frac{1}{2}e^a\boldsymbol{\Gamma}_a\psi_r - A_r^s\psi_s + i\left(\frac{1}{4} - \frac{1}{\mathcal{N}}\right)b\psi_r. \quad (9.13)$$

where D is the covariant derivative in the Lorentz connection.

The above relation implies that the fermions carry a $u(1)$ "electric" charge $q = (\frac{1}{4} - \frac{1}{\mathcal{N}})$. The purely gravitational part L_G is equal to the standard Einstein-Hilbert action with cosmological constant, plus the dimensionally continued Euler density[2].

The action is by construction quasi-invariant under local supersymmetry transformations $\delta_\lambda \boldsymbol{A} = d\lambda + [\boldsymbol{A}, \lambda]$ with $\lambda = \bar{\boldsymbol{Q}}^r\epsilon_r - \bar{\epsilon}^r\boldsymbol{Q}_r$,

$$\delta e^a = \frac{1}{2}\left(\bar{\epsilon}^r\boldsymbol{\Gamma}^a\psi_r - \bar{\psi}^r\boldsymbol{\Gamma}^a\epsilon_r\right),$$

$$\delta\omega^{ab} = -\frac{1}{4}\left(\bar{\epsilon}^r\boldsymbol{\Gamma}^{ab}\psi_r - \bar{\psi}^r\boldsymbol{\Gamma}^{ab}\epsilon_r\right),$$

$$\delta A_s^r = -i\left(\bar{\epsilon}^r\psi_s - \bar{\psi}^r\epsilon_s\right),$$

---

[2]The first terms in L_G are the dimensional continuations of the Euler (or Gauss-Bonnet) densities from four and two dimensions respectively, just as the three-dimensional Einstein-Hilbert Lagrangian is the continuation of the the two dimensional Euler density. The Gauss-Bonnet term is the leading contribution in the limit of vanishing cosmological constant ($l \to \infty$), whose local supersymmetric extension yields a nontrivial extension of the Poincaré group [3].

$$\delta\psi_r = -\nabla\epsilon_r,$$

$$\delta\bar{\psi}^r = -\nabla\bar{\epsilon}^r,$$

$$\delta b = -i\left(\bar{\epsilon}^r\psi_r - \bar{\psi}^r\epsilon_r\right).$$

As can be seen from (9.12) and (9.13), for $\mathcal{N} = 4$ the $U(1)$ field b looses its kinetic term and decouples from the fermions (the gravitino becomes uncharged with respect to $U(1)$). The only remnant of the interaction with the field b is a dilaton-like coupling with the Nieh-Yan and Pontryagin four forms for the AdS$_5$ and $SU(4)$ groups in $L_{u(1)}$.

In the bosonic sector, for $\mathcal{N} = 4$, the field equation obtained from the variation with respect to b states that the Pontryagin four forms of AdS$_5$ and $SU(\mathcal{N})$ groups are proportional. Consequently, the corresponding Chern classes on any four-dimensional submanifold must be related. Since $\pi_4(SU(4)) = 0$, the above implies that the Pontryagin and the Nieh-Yan numbers must cancel out on any closed four manifold without boundary.

9.1.3 *Eleven-dimensional AdS CS supergravity action for* osp(32|1)

In order to construct a supersymmetric extension of the CS AdS gravity action in eleven dimensions, one can take advantage of the fact that $11 = 3 + 8$ and the Majorana representation for spinors is available (Bott periodicity again). In this case, the smallest supersymmetric extension of the AdS group is given by an orthosymplectic group, to wit, OSP(32|1). The corresponding super algebra denoted by osp(32|1) is spanned by the AdS generators together with a fifth-rank bosonic generator $\boldsymbol{J}_{abcde}$ in addition to the Majorana fermionic generator $\boldsymbol{Q}$. The anticommutator of the latter is given by

$$\{\boldsymbol{Q}, \boldsymbol{Q}\} \sim \Gamma^a \boldsymbol{J}_a + \Gamma^{ab} \boldsymbol{J}_{ab} + \Gamma^{abcde} \boldsymbol{J}_{abcde},$$

and the fundamental connection one-form field, expanded along the osp(32|1) generators, reads

$$\boldsymbol{A} = e^a \boldsymbol{J}_a + \frac{1}{2}\omega^{ab} \boldsymbol{J}_{ab} + \frac{1}{5!}b^{abcde} \boldsymbol{J}_{abcde} + \psi \boldsymbol{Q}, \qquad (9.14)$$

where b^{abcde} is a totally antisymmetric fifth-rank Lorentz tensor one-form. Writing down the associated field strength $\boldsymbol{F}$, the CS action for osp(32|1) denoted by $\mathbf{L}_{11}^{osp(32|1)}$ is such that

$$d\mathbf{L}_{11}^{osp(32|1)} = \mathrm{Str}(\boldsymbol{F}^6),$$

where Str stands for the supertrace in the matrix representation (8.8)–(8.10). Then, the Lagrangian splits into two pieces as

$$\mathbf{L}_{11}^{osp(32|1)}(\boldsymbol{A}) = L_{11}^{sp(32)}(\boldsymbol{\Omega}) + L_\psi(\boldsymbol{\Omega}, \psi), \qquad (9.15)$$

where $\boldsymbol{\Omega} \equiv \frac{1}{2}(e^a \boldsymbol{\Gamma}_a + \frac{1}{2}\omega^{ab}\boldsymbol{\Gamma}_{ab} + \frac{1}{5!}b^{abcde}\boldsymbol{\Gamma}_{abcde})$ is an $sp(32)$ connection [4, 5, 146]. The bosonic part of (9.15) can be written as

$$L_{11}^{sp(32)}(\boldsymbol{\Omega}) = 2^{-6}L_{G\,11}^{AdS}(\omega, e) - \frac{1}{2}L_{T\,11}^{AdS}(\omega, e) + L_{11}^b(b, \omega, e),$$

where $L_{G\,11}^{AdS}$ is the CS form associated to the 12-dimensional Euler density, and $L_{T\,11}^{AdS}$ is the CS form whose exterior derivative is the Pontryagin twelve-form for $SO(10, 2)$. The fermionic Lagrangian is

$$\begin{aligned}
L_\psi = {}& 6(\bar\psi \mathcal{R}^4 D\psi) - 3\left[(D\bar\psi D\psi) + (\bar\psi \mathcal{R}\psi)\right](\bar\psi \mathcal{R}^2 D\psi) \\
& -3\left[(\bar\psi \mathcal{R}^3 \psi) + (D\bar\psi \mathcal{R}^2 D\psi)\right](\bar\psi D\psi) \\
& +2\left[(D\bar\psi D\psi)^2 + (\bar\psi \mathcal{R}\psi)^2 + (\bar\psi \mathcal{R}\psi)(D\bar\psi D\psi)\right](\bar\psi D\psi),
\end{aligned}$$

where $\mathcal{R} = d\boldsymbol{\Omega} + \boldsymbol{\Omega}^2$ is the $sp(32)$ curvature. It is a matter of check to see that the supersymmetry transformations read off as a gauge transformations and given by

$$\delta e^a = \frac{1}{8}\bar\epsilon \boldsymbol{\Gamma}^a \psi, \quad \delta\omega^{ab} = -\frac{1}{8}\bar\epsilon \boldsymbol{\Gamma}^{ab}\psi,$$

$$\delta\psi = D\epsilon, \quad \delta b^{abcde} = \frac{1}{8}\bar\epsilon \boldsymbol{\Gamma}^{abcde}\psi,$$

leaving the action quasi-invariant.

Having found this eleven-dimensional supergravity action, it is natural to look for the differences with the standard eleven-dimensional (CJS) supergravity of Cremmer, Julia and Sherk [141]. The CJS theory is an $\mathcal{N} = 1$ supersymmetric extension of Einstein-Hilbert gravity which, as shown in Refs. [142] and [143], does not admit a cosmological constant[3] and its possible $\mathcal{N} > 1$ extensions are not known.

In our case, on the other hand, the cosmological constant is necessarily nonzero by construction. The extension for $\mathcal{N} > 1$ simply requires including an internal $so(\mathcal{N})$ gauge field coupled to the fermions and the resulting Lagrangian is an $osp(32|\mathcal{N})$ CS form [146]. As shown in the next section and in the next chapter, eleven-dimensional CS supergravity action without cosmological constant can be constructed for the M-algebra but in this case, the gravity sector does not contain the Einstein-Hilbert action but only the Poincaré-invariant term (5.50).

Another important difference between the two supergravity actions in eleven dimensions are the field contents of both theories. Indeed, in the standard CJS action, a bosonic three-form field is required which is not present in the CS

---

[3]What the authors of [142, 143] have shown is that a perturbative deformation of the CJS theory to include a nonvanishing (negative) cosmological constant would be inconsistent. The action defined by (9.15) circumvents this obstruction because it is not continuously connected to the $\Lambda = 0$ case: the algebra becomes non-semisimple, the invariant tensor degenerates, the nice matrix representation does not exist, and the whole construction breaks down for $\Lambda = 0$.

construction because the connection one-form in the CS action cannot accommodate a fundamental three-form field without breaking the fiber bundle structure.

9.2 Poincaré CS Supergravity Actions

We now consider Poincaré CS supergravity actions defined in $D = 2n + 1$ dimensions. More precisely, we are going to construct the supersymmetric extension of the Lovelock action,

$$I_P = \int \epsilon_{a_1 \cdots a_{2n+1}} R^{a_1 a_2} \cdots R^{a_{2n-1} a_{2n}} e^{a_{2n+1}} = \int \mathcal{L}_P, \qquad (9.16)$$

which is invariant under the local Poincaré transformations (5.50). The resulting supersymmetric action should produce a genuine CS gauge theory for some extension of the Poincaré group. In order to achieve this, we first consider the generic case with Dirac spinors, and then focus our attention in the particular dimensions $D = 3 + 8k$ where a spinorial Majorana representation is allowed for signature $(1, D - 1)$.

In contrast with the AdS case, the Poincaré algebra is not semisimple, which means that there is no nondegenerate matrix representation like that in terms of Γs. Hence, there is no easy way to construct the invariant tensor and, as explained in the introduction of this chapter, we will opt for a different constructive approach. This is done in various steps in a way analogous to what is known as the Noether procedure — a perturbative scheme to construct all possible consistent interactions terms starting from a Lagrangian for a free theory. If the iterative process stops after a finite number of steps, one ends up with an action principle invariant under a given symmetry group.

As a first step of our construction, we extend the Poincaré connection to a super connection in order to include the fermions on the same footing as the bosonic fields,

$$\boldsymbol{A} = e^a \boldsymbol{P}_a + \frac{1}{2} \omega^{ab} \boldsymbol{J}_{ab} + \bar{\psi}_\alpha \boldsymbol{Q}^\alpha - \bar{\boldsymbol{Q}}_\alpha \psi^\alpha, \qquad (9.17)$$

where the $\boldsymbol{Q}^\alpha$ are the fermionic generators and ψ^α is a Dirac spinor one-form, the *Rarita-Schwinger field*. These generators correspond to the simplest supersymmetric extension of the Poincaré algebra — the *standard super Poincaré algebra*. Its generators satisfy the following (anti)commutation relations

$$[\boldsymbol{P}_a, \boldsymbol{P}_b] = 0, \quad [\boldsymbol{J}_{ab}, \boldsymbol{P}_c] = \boldsymbol{P}_a \eta_{bc} - \boldsymbol{P}_b \eta_{ac}, \quad [\boldsymbol{J}_{ab}, \boldsymbol{J}_{cd}] = -\boldsymbol{J}_{ac} \eta_{bd} + \cdots$$

$$[\boldsymbol{P}_a, \boldsymbol{Q}^\alpha] = 0, \quad [\boldsymbol{J}_{ab}, \boldsymbol{Q}^\alpha] = \frac{1}{2}(\Gamma_{ab})^\alpha_\beta \boldsymbol{Q}^\beta, \quad \{\boldsymbol{Q}^\alpha, \bar{\boldsymbol{Q}}_\beta\} = -i(\Gamma^a)^\alpha_\beta \boldsymbol{P}_a. \qquad (9.18)$$

As shown below, this is sufficient in three dimensions to construct a CS action for this algebra, while for odd dimensions $D \geq 5$, an additional fifth rank bosonic generator will be required in order for the procedure to close. The resulting super algebra with this extra fifth rank generator is referred to as a *five-brane super algebra*. For $D = 5$, the extra generator reduces to a central charge and the algebra acquires a central extension.

From the super Poincaré connection (9.17) and the standard super Poincaré algebra (9.18), we read off the supersymmetric gauge transformations of the dynamical fields with spinor zero-form parameter ϵ as

$$\delta e^a = -i\left(\bar{\epsilon}\Gamma^a\psi - \bar{\psi}\Gamma^a\epsilon\right), \quad \delta\omega^{ab} = 0, \quad \delta\psi = D\epsilon. \tag{9.19}$$

Now, the variation of the action I_P with respect to these transformations requires the introduction of a compensating term involving the Rarita-Schwinger field,

$$I_\psi = \frac{i}{3}\int \epsilon_{abca_1\cdots a_{2n-2}} R^{a_1 a_2}\cdots R^{a_{2n-3}a_{2n-2}}\left(\bar{\psi}\Gamma^{abc}D\psi + \text{h.c.}\right), \tag{9.20}$$

in order to cancel part of the variation of I_P. Note that the variation of this Rarita-Schwinger term, apart from canceling the variation of the Poincaré gravity action, also generates an extra piece given by

$$\delta\left(I_P + I_\psi\right) = \int \epsilon_{abca_1\cdots a_{2n-2}} R^{a_1 a_2}\cdots R^{a_{2n-3}a_{2n-2}} R_{de}\left(\bar{\epsilon}\Gamma^{abcde}\psi - \text{h.c.}\right).$$

It is clear that in three dimensions, this extra piece will be absent and the supersymmetric extension of the Poincaré gravity action reduces to $I_P + I_\psi$. For odd dimensions $D \geq 5$, the presence of this extra fifth-rank 1-form bosonic field means that a possible candidate for the super algebra that ensures a supersymmetric extension of I_P could be given by the super five-brane Poincaré algebra spanned by the following generators

$$\boldsymbol{G}_A = \left[\boldsymbol{P}_a, \boldsymbol{J}_{ab}, \boldsymbol{K}_{abcde}, \boldsymbol{Q}^\alpha, \bar{\boldsymbol{Q}}_\alpha\right], \tag{9.21}$$

where $\boldsymbol{K}_{abcde}$ is the fifth-rank bosonic generator, which becomes a truly central charge only in five dimensions. The anti-commutator of fermionic generators acquires an extra piece proportional to $\boldsymbol{K}_{abcde}$,

$$\left\{\boldsymbol{Q}^\alpha, \bar{\boldsymbol{Q}}_\beta\right\} = -i\left(\Gamma^a\right)^\alpha{}_\beta \boldsymbol{P}_a + i\left(\Gamma^{abcde}\right)^\alpha{}_\beta \boldsymbol{K}_{abcde}. \tag{9.22}$$

One the other hand, the super-Poincaré connection (9.17) has to be extended in order to accommodate this extra generator, and a bosonic fifth-rank field b^{abcde} must also to be added,

$$\boldsymbol{A} \rightarrow \boldsymbol{A} + b^{abcde}\boldsymbol{K}_{abcde}.$$

This in turn implies that the supersymmetric transformations (9.19) are supplemented with the following relation

$$\delta b^{abcde} = -i\left(\bar{\epsilon}\Gamma^{abcde}\psi - \bar{\psi}\Gamma^{abcde}\epsilon\right). \tag{9.23}$$

Finally, it is a simple matter to check that the action defined by

$$I_P + I_\psi + \frac{1}{6}\int \epsilon_{abca_1\cdots a_{2n-2}} R^{a_1 a_2}\cdots R^{a_{2n-3}a_{2n-2}} R_{de}b^{abcde}, \tag{9.24}$$

is invariant under the supersymmetric transformations (9.19) and (9.23).

It is important to stress that the supersymmetry is realized off-shell, i.e., without using the field equations, and the supersymmetric action (9.24) is a CS action for the super five brane Poincaré algebra. Now, we are in a position to write down the invariant tensor for the super five brane Poincaré algebra that yields the supersymmetric action (9.24). The invariant tensor is a $(n+1)$ multilinear form denoted by $< \cdots >$ whose only non-vanishing components are given by

$$< \boldsymbol{J}_{a_1 a_2}, \ldots, \boldsymbol{J}_{a_{2n-1} a_{2n}}, \boldsymbol{P}_{a_{2n+1}} > = \epsilon_{a_1 \cdots a_{2n+1}},$$

$$< \boldsymbol{J}_{a_1 a_2}, \ldots, \boldsymbol{J}_{a_{2n-3} a_{2n-2}}, \boldsymbol{J}_{fg}, \boldsymbol{K}_{abcde} > = -\frac{1}{12} \epsilon_{a_1 \cdots a_{2n-2} abc} \eta_{[fg][de]},$$

$$< \boldsymbol{Q}, \boldsymbol{J}_{a_1 a_2}, \ldots, \boldsymbol{J}_{a_{2n-3} a_{2n-2}}, \bar{\boldsymbol{Q}} > = 2i^n \Gamma_{a_1 \cdots a_{2n-2}}.$$

9.2.1 *Poincaré CS supergravity with Majorana spinors*

We now restrict our analysis to those odd dimensions that allow Majorana spinors, that is $D = 3 + 8k$ with $k \in \mathbb{N}$. We rewrite the gravitational Lagrangian $\mathcal{L}_P$ defined in (9.16) by using a trace expression over the $\Gamma-$matrices as

$$\mathcal{L}_P = \alpha_0 \mathrm{Tr} \left[\rlap{/}{R}^{4k+1} \rlap{/}{e} \right], \tag{9.25}$$

where α_0 is a normalization constant that can be dropped in this discussion, and we have defined

$$\rlap{/}{e} = e_a \Gamma^a, \qquad \rlap{/}{R} = \frac{1}{2} R_{ab} \Gamma^{ab}. \tag{9.26}$$

As previously done, in order to construct a local supersymmetric extension of $\mathcal{L}_P$, the additional fields required by supersymmetry should combine into a connection for some supersymmetric extension of the Poincaré algebra. In this way, the supersymmetric extension of the Poincaré algebra naturally emerges, accommodating the extra fields required by supersymmetry and prescribing their correct supersymmetry transformations. In the simplest (tentative) super-Poincaré theory, the field content is just supplemented by the introduction of a single gravitino ψ, which produces an $\mathcal{N} = 1$ supersymmetric theory. The local supersymmetric transformations, with parameter $\lambda = \bar{\epsilon}^\alpha Q_\alpha$ acting on the fields, take the form

$$\delta e^a = (\bar{\epsilon} \Gamma^a \psi), \quad \delta \omega^{ab} = 0, \quad \delta \psi = D\epsilon := \left(d + \frac{1}{4} \omega_{ab} \Gamma^{ab} \right) \epsilon,$$

where ϵ is a zero-form Majorana spinor. In this case, the variation of $\mathcal{L}_P$ under these supersymmetric transformations can be canceled by a kinetic term of the gravitino ψ given by

$$\mathcal{L}_\psi = -2^{k+1} \mathrm{Tr} \left[\rlap{/}{R}^{4k} (D\psi) \bar{\psi} \right].$$

The combined transformation for the two terms yields

$$\delta\mathcal{L}_P + \delta\mathcal{L}_\psi = \text{Tr}\left[\not{R}^{4k+1}\left((\bar{\epsilon}\Gamma_a\psi)\Gamma^a - 2^k(\epsilon\bar{\psi} - \psi\bar{\epsilon})\right)\right].$$

This expression can be rearranged using the following Fierz identity valid in $D = 3 + 8k$

$$\epsilon\bar{\psi} - \psi\bar{\epsilon} = \frac{1}{2^k}(\bar{\epsilon}\Gamma_a\psi)\Gamma^a + \sum_{p\in\mathcal{P}}\frac{(-1)^{p+1}}{2^k p!}\left(\bar{\epsilon}\Gamma_{a_1\cdots a_p}\psi\right)\Gamma^{a_1\cdots a_p} \qquad (9.27)$$

where the sum is over the set $\mathcal{P}$ defined by

$$\mathcal{P} = \{p = 2, 5 \ (\text{mod } 4) \quad \text{with} \quad p \leq 4k+1\}. \qquad (9.28)$$

Thus, one finally obtains

$$\delta\mathcal{L}_P + \delta\mathcal{L}_\psi = -\sum_{p\in\mathcal{P}}\frac{(-1)^{p+1}}{p!}\text{Tr}\left[\not{R}^{4k+1}\left(\bar{\epsilon}\Gamma_{a_1\cdots a_p}\psi\Gamma^{a_1\cdots a_p}\right)\right], \qquad (9.29)$$

which does not obviously vanish for $D \geq 5$. As before, we conclude that the standard super-Poincaré algebra is not rich enough to ensure the off-shell supersymmetry of the Lagrangian $\mathcal{L}_P$. Nevertheless, it is simple to see that the variation (9.29) can be canceled by introducing bosonic one-form fields that are tensors of rank p, $b_{(p)}^{a_1\cdots a_p}$ with $p \in \mathcal{P}$, that transform as $\bar{\epsilon}\Gamma_{a_1\cdots a_p}\psi$ under supersymmetry. Assuming that these extra fields belong to a single connection, the obvious option is to consider an extension of the Poincaré algebra spanned by the generators

$$\boldsymbol{G}_A = \boldsymbol{J}_{ab}, \boldsymbol{P}_a, \boldsymbol{Q}_\alpha, (\boldsymbol{Z}_{a_1\cdots a_p})_{p\in\mathcal{P}}, \qquad (9.30)$$

where $\boldsymbol{Q}_\alpha$ is the Majorana generator and the generators $(\boldsymbol{Z}_{a_1\cdots a_p})_{p\in\mathcal{P}}$ are Lorentz tensors of rank p. In this case, the corresponding super connection takes the form

$$\boldsymbol{A} = \frac{1}{2}\omega^{ab}\boldsymbol{J}_{ab} + e^a\boldsymbol{P}_a + \psi^\alpha Q_\alpha + \sum_{p\in\mathcal{P}}\frac{1}{p!}b_{(p)}^{a_1\cdots a_p}\boldsymbol{Z}_{a_1\cdots a_p}. \qquad (9.31)$$

In addition, in order to prescribe the correct gauge supersymmetric transformations of the extra bosonic fields, the anticommutator of the Majorana generators must be given by

$$\{\boldsymbol{Q}, \boldsymbol{Q}\} = (C\Gamma^a)\boldsymbol{P}_a + \sum_{p\in\mathcal{P}}\frac{1}{p!}(C\Gamma^{a_1\cdots a_p})\boldsymbol{Z}_{a_1\cdots a_p}, \qquad (9.32)$$

where C is the antisymmetric charge conjugation matrix. The algebra (9.32) is known as the $\mathcal{N} = 1$ *maximal extension of the super-Poincaré algebra*. This algebra is said maximal because the left hand side is a $2^{[D/2]} \times 2^{[D/2]}$ real symmetric matrix, so the maximal number of algebraically distinct charges that can appear on the right hand side is $2^{[D/2]} \times (2^{[D/2]} + 1)/2$, which is precisely the number of components of $\boldsymbol{P}_a$ and the different p-form "central charges" $\boldsymbol{Z}_{a_1\cdots a_p}$ that appear in the right hand side. In eleven dimensions, this algebra is commonly known as the M-algebra since it encodes many important features of the M-theory [147].

The supersymmetry transformations of all the dynamical fields can be read off as a gauge transformation of the connection (9.31) for the algebra (9.32), and they are given by

$$\delta e^a = (\bar{\epsilon}\Gamma^a\psi), \quad \delta\psi = D\epsilon$$

$$\delta\omega^{ab} = 0, \quad \delta b_{(p)}^{a_1\cdots a_p} = (\bar{\epsilon}\Gamma^{a_1\cdots a_p}\psi). \tag{9.33}$$

Finally, the local supersymmetric extension of the Poincaré invariant gravity Lagrangian in the odd dimensions $D = 3 + 8k$ is found to be

$$\mathcal{L}_P^{\text{susy}} = \text{Tr}\Big[\mathcal{R}^{4k}\Big(\mathcal{R}\,\big(\not{e} + \sum_{p\in\mathcal{P}}(-1)^{p+1}\not{b}_{(p)}\big) - (D\psi)\bar{\psi}\Big)\Big], \tag{9.34}$$

where we have defined

$$\not{b}_{(p)} = \frac{1}{p!}b_{(p)}^{a_1\cdots a_p}\Gamma_{a_1\cdots a_p}. \tag{9.35}$$

The invariance of (9.34) under the supersymmetry transformations (9.33) can be easily checked using of the Fierz identity (9.27). Hence, in all odd dimensions that admit Majorana spinors, a supersymmetric extension of the Poincaré invariant gravity can be constructed along these lines and the resulting action is a gauge theory for the maximal extension of the $\mathcal{N} = 1$ super-Poincaré algebra.

Chapter 10

Inönü-Wigner Contractions
and Its Extensions

The Poincaré group is the symmetry of the spacetime that best approximates the world around us at low energy, Minkowski space. The Poincaré group can be viewed as the limit of vanishing cosmological constant or infinite radius ($\Lambda \sim \pm l^{-2} \to 0$) of the de Sitter or anti-de Sitter groups. This deformation is called a *Inönü-Wigner* (**IW**) *contraction* [148] that can be implemented through a rescaling of the generators in the algebra:

$$\mathbf{J}_a \to \mathbf{P}_a = l^{-1}\mathbf{J}_a, \quad \mathbf{J}_{ab} \to \mathbf{J}_{ab}. \tag{10.1}$$

Thus, starting from the AdS symmetry in 3+1 dimensions ($SO(3,2)$), the rescaled algebra is

$$[\mathbf{P}_a, \mathbf{P}_b] = l^{-2}\mathbf{J}_{ab} \tag{10.2}$$

$$[\mathbf{J}_{ab}, \mathbf{P}_c] = \mathbf{P}_a\eta_{bc} - \mathbf{P}_b\eta_{ac} \tag{10.3}$$

$$[\mathbf{J}_{ab}, \mathbf{J}_{cd}] \sim \mathbf{J}_{ad}\eta_{bc} - \cdots . \tag{10.4}$$

Note that in the limit, $l \to \infty$ the right hand side of (10.2) vanishes and $\mathbf{P}_a$ becomes a generator of translations in the Poincaré group. This is defined as the IW contraction $SO(3,2) \to ISO(3,1)$. A similar contraction takes the de Sitter group into Poincaré, or in general, $SO(p,q) \to ISO(p,q-1)$.

In general, a IW contraction is a singular limit that drastically changes the structure constants and the Killing metric of the algebra without changing the number of generators in a way that the resulting algebra is still a Lie algebra. Since some structure constants may vanish under the contraction, as in the example above, some generators become commuting and end up forming an abelian subalgebra. Therefore, the contraction of a semisimple algebra is not necessarily semisimple. For a detailed discussion of contractions, see, e.g., [68], and for a nice historical note, see Ref. [69].

As could be expected, the contraction of a group induces a contraction of representations and therefore it is possible to obtain a Lagrangian for the contracted group by a corresponding limiting procedure. Conversely, as noted by Segal long ago, if two physical theories are linked through a limiting process then there should also exist a corresponding limit between their underlying symmetry groups [149]. In view of this, Lie algebras related by IW contractions and the obtention of new Lie algebras from a given one has become a problem of interest in physics and mathematics. On the other hand, as immediately noticed by its inventors, the IW contractions can give rise to unfaithful representations. In other words, the limit representation may not be an irreducible faithful representation of the contracted group. Therefore, the procedure to obtain an action for the contracted group is not the straightforward limit of the original action. This is particularly delicate in the case of supersymmetric actions. In fact, the CS action for the super-Poincaré group obtained in Ref. [3] is not the naive limit of the Chern-Simons action for super-AdS found in Ref. [4]. We will come to this point in detail later. However, the CS action for the Poincaré group (without supersymmetry) can be easily obtained from the AdS CS gravity action through the simple IW contraction in arbitrary odd dimensions.

For the sake of completeness, we should also mention that there exist various approaches in the literature to obtain a Lie algebra from another one in a way similar to the Inönü-Wigner contraction. In some cases, these procedures are restrictive in the sense that the starting and resulting algebras have the same dimension. In Ref. [150], de Azcárraga *et al.* have proposed a consistent way of generating a Lie algebra whose dimension is greater than the original one. This method, originally considered by Hatsuda and Sakaguchi [151] in a particular context, consists of expanding the Maurer-Cartan one-forms in powers of a real parameter in such way that the Maurer-Cartan equations are satisfied order by order, leading to a closed algebra at each order. Using this expansion process, the authors of Ref. [150] have derived the M-algebra from the superalgebra $osp(32|1)$ which has 55 less generators.

Thus, the M-algebra and the orthosymplectic superalgebra $osp(32|1)$ (see Chapter 8) which arise naturally in the context of eleven-dimensional Chern-Simons supergravity theories are linked through this expansion method. In the spirit of Segal [149], we will see that there exists a limiting process compatible with the expansion that relates the $osp(32|1)$ supergravity theory to the CS supergravity for the M-algebra. More precisely, this process allows to obtain the supergravity action and the supersymmetry transformations for the M-algebra [50] from those of the $osp(32|1)$ supergravity. Note that since these theories have different field content, the contact between them at the level of the actions is not that simple and requires the addition of a Lorentz tensor to the conventional spin connection.

As mentioned before, the Inönü-Wigner contraction is the natural option to take the vanishing cosmological constant limit of a given theory. However, as we will see, the leading terms of the contraction of the $osp(32|1)$ supergravity yields an action decoupled from the vielbein which is of little physical interest while the next to leading terms which contains the vielbein are not supersymmetric. As we will show, this obstruction is due to the presence of the Pontryagin-Chern-Simons form in the original action. This form, which is only defined in $d = 4k - 1$, is required in order to obtain supersymmetric extension of the AdS connection without duplicating the field content of the theory. Hence, the need to expand the osp(32|1) algebra rather than taking its Inönü-Wigner contraction is related to the existence of the Pontryagin invariant in 12 dimensions.

Next, we consider the three-dimensional case which is the simplest and whose supersymmetric extensions can be obtained with or without the torsional terms. We then extend the results to eleven dimensions where we show the way to obtain the CS supergravity actions for the M-algebra from the osp(32|1) supergravity theory.

10.1   Three-Dimensional Case

The three-dimensional case is very instructive and will be useful to extend the procedure to eleven dimensions. As we already mentioned, since there are two characteristic classes in four dimensions (the Euler and Pontryagin forms), there are two CS gravities for the AdS group in three dimensions. Correspondingly, there are two different supergravity theories in three dimensions which are gauge theories for osp(2|1). The Lagrangian of the first one does not involve torsion explicitly, while the second has a piece proportional to the torsion. We will show that for the first theory, the standard IW contraction leads to the $\mathcal{N} = 1$ supersymmetric extension of the three-dimensional Einstein-Hilbert action while the second case requires a generalization of the IW contraction.

10.1.1   *Case without torsion in the action*

The supersymmetric extension of the three-dimensional Einstein-Hilbert Lagrangian with negative cosmological constant is given by

$$L_{G3} = \epsilon_{abc}\left(R^{ab} + \frac{1}{3}e^a e^b\right)e^c - 4\bar{\psi}\nabla\psi, \tag{10.5}$$

where we have absorbed the radius l in the vielbein and the spin covariant derivative is defined by

$$\nabla\psi := \left(d + \frac{1}{4}\omega_{ab}\Gamma^{ab} + \frac{1}{2}e_a\Gamma^a\right)\psi = D(\omega)\psi + \frac{1}{2}e_a\Gamma^a\psi. \tag{10.6}$$

The supersymmetry transformations are

$$\delta e^a = 2\bar{\epsilon}\Gamma^a\psi, \quad \delta\omega^{ab} = -2\left(\bar{\epsilon}\Gamma^{ab}\psi\right), \quad \delta\psi = \nabla\epsilon, \tag{10.7}$$

which can be read off from a gauge transformation assuming that the dynamical fields belong to the connection in the osp(2|1) algebra,

$$\boldsymbol{A} = e^a \boldsymbol{J}_a + \frac{1}{2}\omega^{ab}\boldsymbol{J}_{ab} + \psi^\alpha \boldsymbol{Q}_\alpha.$$

The non-vanishing (anti)commutation relations of osp(2|1) are

$$[\boldsymbol{J}_a, \boldsymbol{J}_b] = \boldsymbol{J}_{ab}, \quad [\boldsymbol{J}_{ab}, \boldsymbol{J}_c] = \boldsymbol{J}_a\eta_{bc} - \boldsymbol{J}_b\eta_{ac}, \quad [\boldsymbol{J}_{ab}, \boldsymbol{J}_{cd}] = -\boldsymbol{J}_{ac}\eta_{bd} + \cdots$$

$$[\boldsymbol{J}_{[p]}, \boldsymbol{Q}_\alpha] = \frac{1}{2}\left(\Gamma_{[p]}\right)_\alpha^{\ \beta} \boldsymbol{Q}_\beta, \quad \{\boldsymbol{Q}_\alpha, \boldsymbol{Q}_\beta\} = 2\left(C\Gamma^a\right)_{\alpha\beta}\boldsymbol{J}_a - \left(C\Gamma^{ab}\right)_{\alpha\beta}\boldsymbol{J}_{ab},$$

$$(10.8)$$

where $\boldsymbol{J}_{[p]}$ denotes denotes both AdS generators, $\boldsymbol{J}_{ab}$ and $\boldsymbol{J}_a$.

A standard Wigner-Inönü contraction consists in rescaling the generators as

$$\boldsymbol{J}_a \mapsto l\,\boldsymbol{P}_a, \quad \boldsymbol{J}_{ab} \mapsto \boldsymbol{J}_{ab}, \quad \boldsymbol{Q}_\alpha \mapsto \sqrt{l}\,\boldsymbol{Q}_\alpha, \qquad (10.9)$$

which in turns implies that at the limit as l goes to infinity, the algebra (10.8) becomes the $\mathcal{N} = 1$ super-Poincaré one, i.e.

$$[\boldsymbol{P}_a, \boldsymbol{P}_b] = 0, \quad [\boldsymbol{J}_{ab}, \boldsymbol{P}_c] = \boldsymbol{P}_a\eta_{bc} - \boldsymbol{P}_b\eta_{ac}, \quad [\boldsymbol{J}_{ab}, \boldsymbol{J}_{cd}] = -\boldsymbol{J}_{ac}\eta_{bd} + \cdots$$

$$[\boldsymbol{P}_a, \boldsymbol{Q}_\alpha] = 0, \ [\boldsymbol{J}_{ab}, \boldsymbol{Q}_\alpha] = \frac{1}{2}\left(\Gamma_{ab}\right)_\alpha^{\ \beta} \boldsymbol{Q}_\beta, \ \{\boldsymbol{Q}_\alpha, \boldsymbol{Q}_\beta\} = 2\left(C\Gamma^a\right)_{\alpha\beta}\boldsymbol{P}_a.$$

The shift of the generators (10.9) is equivalent of rescaling the dynamical fields as

$$e^a \mapsto \frac{1}{l}e^a, \quad \omega^{ab} \mapsto \omega^{ab}, \quad \psi \mapsto \frac{1}{\sqrt{l}}\psi. \qquad (10.10)$$

Under this rescaling, L_{G3} expands as a power series in $1/l$,

$$L_{G3} = \frac{1}{l}\left[\epsilon_{abc}R^{ab}e^c - 4\bar{\psi}D(\omega)\psi\right] + \frac{1}{l^2}\left[-2\bar{\psi}e_a\Gamma^a\psi\right] + \frac{1}{l^3}\left[\frac{1}{3}\epsilon_{abc}e^a e^b e^c\right]$$

$$= \frac{1}{l}\mathcal{L}^{(1)} + \frac{1}{l^2}\mathcal{L}^{(2)} + \frac{1}{l^3}\mathcal{L}^{(3)}. \qquad (10.11)$$

The supersymmetry transformations (10.7) are also affected by the rescaling (10.10). More precisely, as the r.h.s. of the transformations always involve a contraction of two spinors, this side goes like l^{-1} which implies that

$$\begin{cases} \delta^{(0)}e^a = 2\bar{\epsilon}\Gamma^a\psi, \ \delta^{(1)}e^a = 0, \\[2mm] \delta^{(0)}\omega^{ab} = 0, \qquad \delta^{(1)}\omega^{ab} = -\frac{2}{l}\left(\bar{\epsilon}\Gamma^{ab}\psi\right), \\[2mm] \delta^{(0)}\psi = D(\omega)\epsilon, \ \delta^{(1)}\psi = \frac{1}{2l}e_a\Gamma^a\epsilon, \end{cases} \qquad (10.12)$$

where the subscripts (0) and (1) mean zeroth and first order in l^{-1}. It is important to note that the transformations $\delta^{(0)}$ are gauge transformations for the $\mathcal{N} = 1$ super-Poincaré group (10.10) once assuming that the fields are components of a single connection for this group.

The variation of the Lagrangian (10.11) under (10.12) is also a series,

$$\delta L_{G3} = \frac{1}{l}\left[\delta^{(0)}\mathcal{L}^{(1)}\right] + \frac{1}{l^2}\left[\delta^{(1)}\mathcal{L}^{(1)} + \delta^{(0)}\mathcal{L}^{(2)}\right] + \cdots, \qquad (10.13)$$

and, because of the quasi-invariance of L_{G3}, each order $1/l^p$ of (10.13) must be the exterior derivative of some function Σ_p. In particular the Lagrangian $\mathcal{L}^{(1)}$ given by (10.11) is just the CS supergravity Lagrangian for the $\mathcal{N} = 1$ super-Poincaré group whose supersymmetry transformations correspond to $\delta^{(0)}$ defined in (10.12).

In this simple exercise, we have shown that the standard IW is enough in order to obtain a supersymmetric action from the osp(2|1) supergravity action. In what follows, we will see how the presence of the torsional terms requires a generalization of the standard IW contraction in order to obtain a physical sensible supersymmetric theory.

10.1.2 *Case with torsion in the action*

A supersymmetric extension of the three-dimensional AdS CS Lagrangian involving a torsion term, quasi-invariant under supersymmetry transformations (10.7), is given by

$$L_{T3} = L_3^\star(\omega) + 2e_a T^a + 8\bar{\psi}\nabla\psi. \tag{10.14}$$

Here $L_3^\star$ is the Lorentz three-form satisfying $dL_3^\star(\omega) = R^{ab}R_{ba}$ and $T^a = de^a + \omega^a{}_b e^b$ is the torsion two-form.

Under the standard Inönü-Wigner contraction (10.10), this Lagrangian expands in powers of $1/l$ as

$$L_{T3} = L_3^\star(\omega) + \frac{1}{l}\left[8\bar{\psi}D(\omega)\psi\right] + \frac{1}{l^2}\left[2e_a T^a + 4\bar{\psi}e_a\Gamma^a\psi\right]$$

$$= \mathcal{L}^{(0)} + \frac{1}{l}\mathcal{L}^{(1)} + \frac{1}{l^2}\mathcal{L}^{(2)} \tag{10.15}$$

and, as in the previous case, its variation at different orders l^{-1} can be written as

$$\delta L_{T3} = \delta^{(0)}\mathcal{L}^{(0)} + \frac{1}{l}\left[\delta^{(0)}\mathcal{L}^{(1)} + \delta^{(1)}\mathcal{L}^{(0)}\right]$$

$$+ \frac{1}{l^2}\left[\delta^{(0)}\mathcal{L}^{(2)} + \delta^{(1)}\mathcal{L}^{(1)}\right] + \cdots \tag{10.16}$$

The first term vanishes identically because $\mathcal{L}^{(0)}$ depends only on the spin connection whose variation $\delta^{(0)}\omega^{ab}$ is zero. Hence, in spite of the fact that the IW contraction at this level gives a consistent quasi-invariant Lagrangian, $\mathcal{L}^{(0)}$ being trivially invariant makes it of little interest. The expression at the next order is more interesting but it involves the transformations acting on two different Lagrangians, which means that it is not obvious how to identify the quasi-invariant Lagrangian. This difficulty is due to the fact that $\mathcal{L}^{(0)}$ depends on the spin connection and $\delta^{(1)}\omega$ is not zero (10.12). A way to circumvent this problem is by splitting the spin connection in two parts with different rescalings, i.e.,

$$\omega^{ab} \mapsto \omega^{ab} + \frac{1}{l}b^{ab}. \tag{10.17}$$

This is compatible with the fact that it is always possible to add to the spin connection a tensor under the Lorentz group. As a direct consequence, the transformations (10.12) now become

$$\begin{cases} \delta^{(0)}e^a = 2\bar{\epsilon}\Gamma^a\psi & \delta^{(1)}e^a = 0 \\ \delta^{(0)}\omega^{ab} = 0 & \delta^{(1)}\omega^{ab} = 0 \\ \delta^{(0)}b^{ab} = -2\left(\bar{\epsilon}\Gamma^{ab}\psi\right) & \delta^{(1)}b^{ab} = 0 \\ \delta^{(0)}\psi = D(\omega)\epsilon & \delta^{(1)}\psi = \frac{1}{2l}e_a\Gamma^a\epsilon \end{cases} \tag{10.18}$$

The series (10.15) is also affected by this splitting of the spin connection and becomes

$$L_{T3} = L_3^\star + \frac{1}{l}\left[8\bar{\psi}D(\omega)\psi + 2R_{ab}\,b^{ba}\right] + o(l^{-2})$$

while its variation now reduces to

$$\delta L_{T3} = \frac{1}{l}\delta^{(0)}\left[8\bar{\psi}D(\omega)\psi + 2R_{ab}\,b^{ba}\right] + o(l^{-2}). \tag{10.19}$$

Hence, the new Lagrangian

$$\mathcal{L}(\omega, b, \psi) = 8\bar{\psi}D(\omega)\psi + 2R_{ab}\,b^{ba}, \tag{10.20}$$

is invariant under the following transformations

$$\delta^{(0)}\omega^{ab} = 0, \quad \delta^{(0)}b^{ab} = -2\left(\bar{\epsilon}\Gamma^{ab}\psi\right), \quad \delta^{(0)}\psi = D(\omega)\epsilon. \tag{10.21}$$

Note that, identifying the one-form b^{ab} as the dual of the vielbein $b^{ab} = \epsilon^{abc}e_c$ in three dimensions, [1] the Lagrangian (10.20) is precisely the $\mathcal{N} = 1$ supersymmetric extension of the three-dimensional Einstein-Hilbert Lagrangian $\mathcal{L}^{(1)}$ in (10.11).

From an algebraic point of view, the splitting (10.17) corresponds to a generalized IW contraction (or expansion) which allows to construct Lie algebras of increasingly higher dimensions from a given one [150]. In our case, the starting algebra was osp(2|1) and the resulting algebra is spanned by the set of generators $G_A = \{J_{ab}, Z_{ab}, Q_\alpha\}$ where Z_{ab} can be identified in three dimensions with the translation generator P_a as $P_a = \epsilon_{abc}Z^{bc}$.

Finally, we mention that there exists an alternative method to derive this result in three dimensions. Indeed, in the expression of order $1/l$ in (10.16), it is easy to show that the variation $\delta^{(1)}\mathcal{L}^{(0)}$ can be written as a $\delta^{(0)}$-variation, i.e.,

$$\delta^{(1)}\mathcal{L}^{(0)} = \left[\delta^{(1)}\omega^{ab}\right]\frac{\delta\mathcal{L}^{(0)}}{\delta\omega_{ab}} = \left[-2(\bar{\epsilon}\Gamma^{ab}\psi)\right]R_{ab}$$

$$= -2\epsilon^{abc}(\delta^{(0)}e_c)R_{ab} = \delta^{(0)}\left[-2\epsilon^{abc}R_{ab}e_c\right],$$

and therefore δL_{T3} reduces to

$$\delta L_{T3} = \frac{1}{l}\delta^{(0)}\left[8\bar{\psi}D(\omega)\psi - 2\epsilon^{abc}R_{ab}e_c\right] + o(l^{-2}),$$

---

[1]This relation is consistent with the supersymmetry transformations, $\delta^{(0)}b^{ab} = \epsilon^{abc}\delta^{(0)}e_c$.

where the expression in brackets can be recognized as the $\mathcal{N} = 1$ supersymmetric extension of the three-dimensional Einstein-Hilbert action at first order in (10.11).

10.2   From osp(32|1) CS Sugra to CS Sugra for the M-Algebra

We now turn to the eleven-dimensional case, which presents a certain physical interest in the context of superstrings and in particular for M-theory. As we have shown, a CS supergravity can be constructed for the osp(32|1) superalgebra, which is presumably relevant for the M-algebra. The latter has 55 more generators than the first and we will show how to relate the two using the expansion method applied at the level of the action, as in the three-dimensional case. Similarly, it will be shown how to derive the super Poincaré transformations from those for the super AdS$_{11}$ algebra [152].

First let us recall that there exists a CS supergravity action for the M-algebra (see previous chapter) whose generators are given by (9.30),

$$\left\{ \boldsymbol{J}_{ab}, \boldsymbol{P}_a, \boldsymbol{Q}_\alpha, (\boldsymbol{Z}_{a_1 \cdots a_p})_{p=2,5} \right\}, \tag{10.22}$$

where $\boldsymbol{Q}_\alpha$ is the Majorana generator and the generators $(\boldsymbol{Z}_{a_1 \cdots a_p})_{p=2,5}$ are Lorentz tensors of second and fifth ranks, and the anticommutator of the fermionic generators reads

$$\{\boldsymbol{Q}, \boldsymbol{Q}\} = (C\Gamma^a)\boldsymbol{P}_a + (C\Gamma^{ab})\boldsymbol{Z}^{(2)}_{ab} + (C\Gamma^{a_1 \cdots a_5})\boldsymbol{Z}^{(5)}_{a_1 \cdots a_5}. \tag{10.23}$$

The supersymmetric Lagrangian for the M-algebra [153] can be simply written as a trace,

$$\mathcal{L}^{\text{susy}}_P = \text{Tr}\left[\cancel{R}^{4k}\left(\cancel{R}\left(\cancel{e} - \cancel{b}_{(2)} + \cancel{b}_{(5)} \right) - (D\psi)\bar{\psi}\right)\right], \tag{10.24}$$

where we have defined

$$\cancel{b}_{(p)} = \frac{1}{p!}b^{a_1 \cdots a_p}_{(p)}\Gamma_{a_1 \cdots a_p}.$$

The invariance of (10.24) under supersymmetry transformations

$$\delta e^a = \bar{\epsilon}\Gamma^a\psi,\ \delta\psi = D\epsilon,\ \delta\omega^{ab} = 0,\ \delta b^{a_1 \cdots a_p}_{(p)} = \bar{\epsilon}\Gamma^{a_1 \cdots a_p}\psi, \tag{10.25}$$

can be easily checked using the following Fierz rearrangement (9.27).

On the other hand, there also exists a CS supergravity action for the osp(32|1) group which is the smallest supersymmetric extension of the AdS algebra in eleven dimension. The supersymmetric Lagrangian is defined as

$$d\mathcal{L}_{\text{osp}(32|1)} = \text{STr}\left[\boldsymbol{F}^6\right], \tag{10.26}$$

where "STr" stands for the supertrace, and $\boldsymbol{F}$ is the osp(32|1) curvature (see Chapters 7 and 8). The supersymmetry transformations for the fields are given by

$$\delta e^a = \bar{\epsilon}\Gamma^a\psi, \quad \delta\psi = \nabla\epsilon$$
$$\delta\omega^{ab} = -\left(\bar{\epsilon}\Gamma^{ab}\psi\right), \quad \delta b^{a_1 \cdots a_5}_{(5)} = \bar{\epsilon}\Gamma^{a_1 \cdots a_5}\psi \tag{10.27}$$

where the covariant derivative now reads

$$\nabla\epsilon = D\epsilon + (e_a\Gamma^a + b^{a_1\cdots a_5}_{(5)}\Gamma_{a_1\cdots a_5})\epsilon.$$

As we are dealing with a theory in presence of a negative cosmological constant Λ, a natural question to ask is whether the limiting case $\Lambda \to 0$ yields an interesting theory. One can see the necessity of considering the expansion method rather than a standard IW contraction by trial and error: consider the standard contraction on the fields,

$$e^a \to \frac{1}{l}e^a, \quad \omega^{ab} \to \omega^{ab}, \quad b^{a_1\cdots a_5}_{(5)} \to \frac{1}{l}b^{a_1\cdots a_5}_{(5)}, \quad \psi \to \frac{1}{\sqrt{l}}\psi, \qquad (10.28)$$

where the zero cosmological constant limit corresponds to taking $l \to \infty$. On the other hand, the gauge parameter ϵ of the supersymmetry transformations must also be rescaled as $\epsilon \to \epsilon/\sqrt{l}$ and the transformations defined by (10.27) reduce to those associated to the extended super-Poincaré algebra (10.25) with the exception that the bosonic field b^{ab} is not present in the AdS theory. Rescaling the supersymmetric action (10.26) with (10.28) gives the expansion

$$\mathcal{L}_{\text{osp}(32|1)} = \mathcal{L}^\star(\omega) + \frac{1}{l}\text{Tr}\left[\mathbb{R}^5\left(\slashed{e} + \slashed{b}_{(5)}\right) - \mathbb{R}^4(D\psi)\bar\psi\right] + o(l^{-2}),$$

$$= \mathcal{L}^{(0)} + \frac{1}{l}\mathcal{L}^{(1)} + o(l^{-2}) \qquad (10.29)$$

where $\mathcal{L}^{(0)} = \mathcal{L}^\star(\omega)$ is the Lorentz Chern-Simons form which depends only on the spin connection,

$$d\mathcal{L}^\star(\omega) = \text{Tr}\left[\mathbb{R}^6\right]. \qquad (10.30)$$

It is clear that in the limit $l \to \infty$, the supersymmetric Lagrangian $\mathcal{L}_{\text{osp}(32|1)}$ reduces to $\mathcal{L}^{(0)}$, which is trivially supersymmetric under (10.25) since $\delta\omega^{ab} = 0$. This means that although the standard IW contraction gives rise to a consistent theory, the resulting Lagrangian decouples from the vielbein and the gravitino, so it does not describe a metric theory for spacetime and is of little interest as a model for gravity. The next order in the expansion (10.29) is more interesting since it contains the Poincaré invariant gravity Lagrangian. However, it is clear that the expression of order l^{-1} cannot be supersymmetric by itself because it lacks the bosonic one-form field $b^{ab}_{(2)}$ needed to close the M-algebra (10.23).

As in three-dimensions, the natural way to circumvent this problem is to exploit the fact that one can always add a Lorentz tensor to the spin connection,

$$\omega^{ab} \to \omega^{ab} - \frac{1}{l}b^{ab}_{(2)}. \qquad (10.31)$$

Apart from introducing the required bosonic field $b^{ab}_{(2)}$, the splitting (10.31) has two other important consequences. First, it prescribes the correct supersymmetric transformation of the bosonic field $b^{ab}_{(2)}$,

$$\delta\omega^{ab} - \frac{1}{l}\delta b^{ab}_{(2)} = -\frac{1}{l}\left(\bar{\epsilon}\Gamma^{ab}\psi\right) \Rightarrow \delta\omega^{ab} = 0, \; \delta b^{ab}_{(2)} = \left(\bar{\epsilon}\Gamma^{ab}\psi\right).$$

Second, the Lorentz CS form $\mathcal{L}^{\star}$ picks up a new contribution of order l^{-1} in the expansion,

$$\mathcal{L}^{\star}(\omega - \frac{1}{l}b_{(2)}) = \mathcal{L}^{\star}(\omega) - \frac{1}{l}\mathrm{Tr}\left[\not{R}^{5}\not{b}_{(2)}\right] + o(l^{-2}).$$

Combining this expression together with (10.29) shows that the order l^{-1} in the expansion gives precisely the supersymmetric action associated with the maximal super-Poincaré algebra (10.24).

So far we have been concerned with the 6th Chern character whose potential CS is given by the Lagrangian $\mathcal{L}_{\mathrm{osp}(32|1)}$, (10.26). However, in eleven dimensions, there exist two more Chern characters, namely, $\mathrm{STr}(\boldsymbol{F}^4)\,\mathrm{STr}(\boldsymbol{F}^2)$ and $\left[\mathrm{STr}(\boldsymbol{F}^2)\right]^3$. A natural question to ask is whether our conclusions depend on the choice of the Chern character. However, after a tedious but straightforward computation, one can realize that any linear combination of all three Chern characters,

$$\alpha_1\mathrm{STr}(\boldsymbol{F}^6) + \alpha_2\mathrm{STr}(\boldsymbol{F}^4)\,\mathrm{STr}(\boldsymbol{F}^2) + \alpha_3\left[\mathrm{STr}(F^2)\right]^3,$$

leads through this procedure to the same conclusion: the IW contraction gives a zeroth-order Lagrangian decoupled from the vielbein. In the next order the maximal extension of the $\mathcal{N} = 1$ super-Poincaré algebra naturally emerges. The resulting Lagrangian is the one derived in the previous section (10.24) up to some additional terms decoupled from the vielbein that are supersymmetric by themselves.

In this concrete example, the need to expand the minimal super AdS algebra rather than performing a standard IW contraction is a consequence of the presence of the Lorentz-CS form (10.30). On the other hand, in five dimensions this form is absent and it is easy to see that the conventional IW contraction works directly. Indeed, the standard scaling (10.28) with b^{abcde} playing the role of $u(1)$ field, $b^{abcde} \propto \epsilon^{abcde}b$, together with the $su(\mathcal{N})$ scaling $A^r_s \to \frac{1}{l}A^r_s$ applied to the $su(2,2|\mathcal{N})$ action (9.12) yields the super Poincaré action (9.24) with a central charge at first order in l^{-1}.

Finally, we mention that this construction is valid in all odd dimensions that admit Majorana spinors, which is $D = 3+8k$. In this case, the maximal extension of the $\mathcal{N} = 1$ super-Poincaré algebra has $D(D-1)/2$ more generators than the minimal supersymmetric extension of the AdS algebra. The expansion method presented here allows to connect the CS supergravity action for the $\mathcal{N} = 1$ super-Poincaré algebra with the CS supergravity action for the minimal extension of the AdS algebra.

Chapter 11

Unconventional Supersymmetries

The fact that SUSY is not manifest in the Standard Model is interpreted as being broken — spontaneously or otherwise — at high energies. Each time a new search fails to find traces of its presence, it is taken as an indication that the breaking must occur at some even higher energy. Thus, if SUSY does not exist we will never know because the supersymmetry breaking scale can always be pushed higher. This makes the SUSY hypothesis impossible to falsify, which is a fundamental criterion for any claim of scientific value [154].

One of the expected signals of standard SUSY is the existence of partners that duplicate the spectrum of observed particles [126]. In the minimal supersymmetric scenario ($\mathcal{N} = 1$ SUSY), for every lepton, quark and gauge quantum, a corresponding particle/field with identical quantum numbers but differing by $\hbar/2$ in intrinsic angular momentum would exist. So far no pairs of particles that even approximately reflect this degeneracy have been observed.

As we saw in Chapter 9, if supersymmetry is in the adjoint representation and the fields are components of a connection for the superalgebra, there is no matching between fermionic and bosonic fields. Those fields seem to respect the pattern of the Standard Model, where bosons are responsible for the gauge interactions and fermions are the constituents matter and make the conserved currents that act as sources for the gauge fields. Can this idea be used to construct more realistic particle models? In Ref. [155] it is shown how to achieve this in $2 + 1$ dimensions and in Ref. [156] it is discussed how the construction can be extended to higher dimensions. In what follows we review the essential features of the construction and refer to the original articles for the details.

11.1 Local Supersymmetry without Gravitini

The origin of the mass degeneracy can be traced back to the assumption in standard (global) SUSY that all fundamental fields are in a vector representation of the supercharge Q. Since SUSY is usually expected to be defined in a globally flat Poincaré-invariant spacetime, Q commutes with the Hamiltonian and therefore all states related by supersymmetry must have the same energy. However, the lack of evidence for supersymmetry in the experiments is not due to the curvature of spacetime.

At cosmic scales, assuming spacetime to be flat is clearly unrealistic in view of the fact that we live in an evolving spacetime that need not possess a particular symmetry anywhere at any given time. Poincaré symmetry, however, is a very good approximation at the microscopic level. In order to see some effect of the spacetime curvature in quantum mechanics, the local radius of curvature ρ should be comparable with the Compton wavelength of particles λ_c, which is certainly not the case: for an electron near the surface of a neutron star $\lambda_c \sim 10^{-15}\rho$. If spacetime is not flat the supersymmetry generators need not commute with the Hamiltonian, but the correction to the mass degeneracy can be estimated to be of order $\delta m \sim \hbar/(c\rho)$.

Hence, the absence of superpartners for the known particles cannot be attributed to a deviation from the Poincaré algebra that corrects supersymmetry. As we saw in Chapter 9, if the fields are in an *adjoint representation* of the superalgebra as parts of a connection, the SUSY partners no longer occur in degenerate pairs. In what follows, we consider yet another representation, keeping the essence of the supersymmetric paradigm — fermions and bosons combined under a supergroup. The main difference with the previous case is that there are no fundamental spin-3/2 fields, and this leads to models that deviate from standard SUSY in some important aspects:

- Supersymmetry is an extension of the local Lorentz (not Poincaré) symmetry.
- The fundamental fields are 1-forms spin-1 bosons and 0-form spin-1/2 fermions, all combined into a connection one-form for a superalgebra. Fields do not belong to vector representations of the superalgebra.
- Only spin-1/2 and spin-1 fields: no fundamental gravitini or scalars, the gravitino and the metric are derived as composite objects.
- The theory is defined in a curved, dynamical spacetime background: there is gravitation, but the metric is invariant under supersymmetry.

11.2   Matter and Interaction Fields

As already emphasized, connection fields responsible for electromagnetic, weak and strong interactions do not require a metric structure. Their transformation laws are independent of the background geometry and it is even possible to define propagating field theories for these gauge connections without mentioning the metric (CS actions). Matter fields, like leptons and quarks, on the other hand, do require a metric. This is because they are fermions represented by zero-form Dirac fields in a particular representation of the Lorentz group that acts on the tangent space (and happens to be the gauge group of the gravitational interaction). Thus fermions necessarily couple to the Lorentz connection and this is ultimately why fermions constitute sources for spacetime geometry, as seen in our discussions of supergravity.

Moreover, in order to define an action principle for a Dirac field it is necessary to introduce the derivative operator $\not{\partial}$ that takes the zero-form spinor ψ into the zero-form $\not{\partial}\psi$ in the same Lorentz representation. This requires an operation that maps the gradient $\partial_\mu \psi$, which is a 1-form on the base manifold, into a zero form spinor in the tangent space. This is achieved by a combination of the Dirac matrices Γ^a and the inverse vielbein E_a^μ. The Dirac matrices are elements of the Clifford algebra associated to the Lorentz group, $\{\Gamma^a, \Gamma^b\} = 2\eta^{ab}$ and are therefore part of the equipment of the tangent space. This projection of the gradient onto the tangent space is given by

$$\not{\partial}\psi \equiv \Gamma^a E_a^\mu \partial_\mu \psi. \tag{11.1}$$

In this way, Dirac's construction ensures that both ψ and $\not{\partial}\psi$ are objects of the same type (zero-form spinors).

The kinetic terms for matter fields in n dimensions can also be expressed as

$$\overline{\psi}\not{\partial}\psi\sqrt{-g}d^n x \equiv \epsilon_{a_1 a_2 \cdots a_n} \overline{\psi} e^{a_1} e^{a_2} \cdots e^{a_{n-1}} \Gamma^{a_n} d\psi, \tag{11.2}$$

where the need for a metric structure in the manifold is manifest.

In conclusion, we arrive at an important difference between matter, represented by spin-1/2 fields, and interactions, represented by connection 1-forms: matter always requires a metric structure, while interactions in odd dimensions do not. On the other hand, since there are no odd-dimensional topological invariants, there are no metric-free counterparts of the CS forms in even dimensions. Therefore, the action for a gauge connection in even dimensions still needs a metric structure. This has important consequences for the implementation of supersymmetry in a model of fundamental spin-1/2 fermions and gauge connections.

11.3   Combining Matter and Interaction Fields

The fact that matter fields transform in a fundamental (spinor) representation of the Lorentz group and of any other gauge group, makes them the ideal bridge between spacetime and internal symmetries. The key ingredient that allows supersymmetry to connect spacetime and internal symmetries is the fact that the supercharge has nontrivial commutators with the internal and spacetime symmetry generators, (8.1)–(8.3).

In what follows, we present a scheme where the only fields are a gauge connection A^k, a Lorentz connection ω^{ab}, a metric structure defined by the vielbein e^a, and a spin-1/2 fermion ψ. The fermion is a zero form and the rest are one-forms. We denote by $\boldsymbol{T}_K$ and $\boldsymbol{J}_{ab}$ the gauge and Lorentz generators. These, together with the supercharges Q and $\overline{Q}$ are assumed to obey a closed superalgebra of the form,

$$[J\ ,\ J] \sim J\ , [T,\ T] \sim T\ , [J,\ T] = 0\ ,$$
$$[J,\ Q] \sim Q\ ,[J,\overline{Q}] \sim -\overline{Q}\ ,$$
$$[T\ ,\ Q] \sim Q\ ,[T,\overline{Q}] \sim -\overline{Q},$$
$$\{Q,\overline{Q}\} \sim T + J\ . \tag{11.3}$$

With these generators the following connection is defined [155, 156]

$$A = A^K T_K + \frac{1}{2}\omega^{ab} J_{ab} + \overline{Q}\not\!\phi\psi + \overline{\psi}\not\!\phi Q + \cdots\ , \tag{11.4}$$

where the metric structure enters through $\not\!\phi = \Gamma_a e^a$, which together with the zero-form spinor ψ combine to produce a connection one-form field *in lieu* of the gravitino. The ellipsis $(\cdots)$ refers to all the additional terms needed to close the superalgebra. In three dimensions there is no need for additional terms [155], but in higher dimensions their presence cannot be ignored [156].

A most important difference between this connection and those used in the preceding chapters is that here the fermionic one-form $\not\!\phi\psi$ is a composite field, while in chapters 9 and 10 it would be a fundamental spin-3/2 field ψ_μ^α, the gravitino. In this scheme spin-3/2 fermions are not fundamental because $\not\!\phi\psi$ belongs to the *reducible representation* $1 \otimes \frac{1}{2} = \frac{1}{2} \oplus \frac{3}{2}$.

With the connection (11.4) one can define an action of the Yang-Mills type in every dimension, or a CS form in odd dimensions. This construction gives rise to models where bosons are massless interaction carriers described by connection fields in the adjoint representation of the gauge algebra, while fermions are possibly massive spinors in the fundamental representations of the gauge group (sections in the gauge bundle), whose currents are sources for the bosonic fields as in the Standard Model.

For example, the action for a model with $SU(2)$ gauge symmetry in $2+1$ dimensions is [157, 158]

$$L_0 = L_{\rm CS}(\omega^a) + L_{\rm CS}(A^K) + d^3x|e|\mathcal{L}_\psi\ , \tag{11.5}$$

where $|e| = \det[e^a{}_\mu]$, $L_{\rm CS}(\omega^a)$ is the Lagrangian of CS gravity, $L_{\rm CS}(A^K)$ is the CS form for the $su(2)$ connection A^K and

$$d^3x|e|\mathcal{L}_\psi = \kappa\ \overline{\psi}[\Gamma^\mu \overrightarrow{D}_\mu - \overleftarrow{D}_\mu\Gamma^\mu + \frac{1}{2}\epsilon_a{}^{bc}T^a{}_{bc}]\psi\ , \tag{11.6}$$

is the fermionic Lagrangian minimally coupled to the $SU(2)$ potential and the Lorentz connection, plus a nonminimal coupling with the torsion. The derivatives $\overrightarrow{D}$ and $\overleftarrow{D}$ stand for

$$\overrightarrow{D} = I_{su(2)} \otimes I_{so(2,1)}d + \frac{1}{2}I_{su(2)} \otimes \omega^a\Gamma_a - \frac{1}{2}A^I\sigma_I \otimes I_{so(2,1)},$$
$$\overleftarrow{D} = \overleftarrow{d}I_{su(2)} \otimes I_{so(2,1)} - \frac{1}{2}I_{su(2)} \otimes \omega^a\Gamma_a + \frac{1}{2}A^I\sigma_I \otimes I_{so(2,1)}. \tag{11.7}$$

In four dimensions, the model with $U(1)$ gauge symmetry constructed using the Yang-Mills form gives the Lagrangian [156]

$$L = [L_F + L_{EM}] \sqrt{|g|} d^4 x - \frac{1}{16} \epsilon_{abcd} \left[R^{ab} - \mu^2 e^a e^b \right] \left[R^{cd} - \mu^2 e^c e^d \right], \qquad (11.8)$$

where $L_{EM} = -\frac{1}{4} F_{\mu\nu} F^{\mu\nu}$, and the fermionic Lagrangian is

$$L_F = \frac{1}{2} \left[\overline{\psi} (\overleftarrow{\nabla} - \overrightarrow{\nabla}) \psi + 4\mu \overline{\psi}\psi \right] + t^\mu \overline{\psi} \Gamma_5 \Gamma_\mu \psi - \frac{1}{3\mu^2} \left[(\overline{\psi}\psi)^2 - (\overline{\psi}\Gamma_5\psi)^2 \right]. \quad (11.9)$$

Here $\overrightarrow{\nabla}\psi = (\slashed{\partial} - i\slashed{A} + \frac{1}{2}\slashed{\omega})\psi$, and $\overline{\psi}\overleftarrow{\nabla} = \overline{\psi}(\overleftarrow{\slashed{\partial}} + i\slashed{A} - \frac{1}{2}\slashed{\omega})$, are the covariant derivatives for the connection of the $U(1) \times SO(3,1)$ gauge group in the spinorial representation, and $t^\mu \equiv -\frac{1}{3!} \varepsilon^{\mu\nu\rho\tau} e^a_\nu T_{a\rho\tau} |e|$.

In this case, in addition to the torsional coupling required by hermiticity, there is a Nambu–Jona-Lasinio term, with a coefficient related to the fermion mass (μ) and the positive cosmological constant (μ^2). The positive cosmological constant here is consonant with the fact that local supersymmetry in this model (and in every even dimension) is necessarily broken, for the same reason discussed earlier, that there are no Lagrangians in even dimensions invariant under local dS, AdS or Poincaré groups. Since the supersymmetry algebra in four dimensions closes on the AdS algebra, the supersymmetry transformations cannot be realized as exact local symmetries for the theory.

An interesting aspect of this construction is its uniqueness: given the gauge group and the spacetime dimension, the connection — and consequently the action — is completely determined. Another attractive feature is that gravitation is essentially required by fermions: the metric structure is needed to define the derivative operator and the connection is required to make the derivative compatible with local Lorentz invariance.

It can be observed that there are no Bose–Fermi pairs. Particles of different spins need not have the same mass and all fields are coupled in the standard gauge-invariant way. In the examples discussed here bosons remain massless while the fermions receive a mass whose origin is in the geometry of spacetime. The theory can be defined in an arbitrarily curved background, and since SUSY requires the inclusion of gravity, this model could be seen as lying in between standard SUSY and SUGRA. In contrast to supergravity, however, all fermions are spin-1/2 particles and no gravitini are included.

Chapter 12

Concluding Remarks

Here we have surveyed a number of theories of gravity in the Cartan first order formulation. These theories have local Lorentz symmetry and are by construction invariant under general coordinate transformations. They can have, however a number of arbitrary dimensionful constants, making the cosmological constant problem seem like a minor complication. These theories coincide with the metric theories of the Lovelock family in the torsion-free sector.

In the odd dimensional case, these actions can be further required to be quasi-invariant under local de Sitter, anti-de Sitter or Poincaré transformations, which is achieved by a selection of the arbitrary coefficients in the action. In this way, there exist three particular choices of coefficients for which these gravitation theories become gauge theories. The miracle is because those particular choices turn the Lagrangian into CS forms for the corresponding Lie algebras.

The supersymmetric extensions of the odd-dimensional theories can also be constructed most easily, owing to their CS nature. The resulting Lagrangians are in this case CS forms for the supersymmetric extensions of, anti-de Sitter or Poincaré algebras in odd dimensions.

The remarkable property of all these theories is that they have no arbitrary adjustable parameters. All the coefficients except for a global factor are fixed, dimensionless, rational numbers. The overall factor in the action turns out to be quantized very much in the same way the electric charge is quantized in the presence of a magnetic monopole.

On the other hand, the lack of adjustable coupling constants makes it very hard to conceive a perturbative scheme to describe interactions. This poses a challenge to the standard quantization schemes where the fields are supposed to be free and behave classically most of the time, with perturbative corrections brought in by quantum corrections. The only natural way to couple CS systems to external sources seems to be by analogy with the minimal coupling between one-form $U(1)$ connection and the world history of a 0-brane in electrodynamics. This idea extended to $(2n + 1)$-CS forms coupled to the world history of a $2n$ brane yields a consistent, gauge invariant model, but its quantum mechanical perturbative expansion is still a formidable open problem.

The irresistible fascination of supersymmetry prompts the extension of the CS gravity theories into a path full of new fields and couplings that are tightly restricted by the symmetry, but which poses an additional puzzle: why is supersymmetry not manifest? How is it broken? Or, as a mystic would say, why Nature does not seem to use it? An answer to this riddle could be hinted at from the discussion of Chapter 11: we might be looking for supersymmetry in a guise that we do not expect, but it might have been sitting there all the time.

One hundred years ago, Einstein presented to the world the equations that, to the best of our knowledge, describe the evolution of the cosmos. This remarkable achievement was the result of a deep examination of the laws of physics and the demand of their consistency. The construction contained the germs of many predictions far beyond the initial scope of the theory, which led to new puzzles: the accelerated expansion of the universe, dark matter, dark energy, the cosmological constant, the delicate imbalance between matter and antimatter, among others.

The quantum description of gravity, possibly the most pressing among the open questions posed by General Relativity, has defied theoretical physicists over most of this century. Merely on statistical grounds, it seems safe to say that if such tenacious struggle by some of the best minds could not defeat the problem, there is a good chance that something is wrong in the way we are used to think about gravitational phenomena. Is the continuum hypothesis about spacetime manifold a reasonable assumption? Could the picture of smooth surface be the right one at a microscopic level or is it just an approximation valid from our coarse-grained macroscopic experience? Could quantum spacetime have a granular, fractal or textile nature? How could we test these possibilities? Is the 3+1 dimensionality of spacetime a fundamental property beyond the Planck scale or is the spacetime dimension an effective feature of our classical experience? These and many other open questions are not addressed here and go beyond the scope of this essay, mainly for want of theoretical tools to handle these issues scientifically. We have merely shown that there are many new interesting and perhaps consistent theories that can be devised to describe the dynamics of the spacetime geometry in D dimensions, and especially so if $D = 2n + 1$.

Ninety years of experience with quantum mechanics has shown that we know how to formulate the quantum theory for simple, nearly linear, weakly coupled systems described by an action principle. But blindly applying "quantization rules" often leads to paradoxes, inconsistencies and dead ends. It is also possible that the efforts to "quantize" gravity suffer from our lack of a deeper understanding of what one means by quantization. Perhaps it is time to think it over.

Bibliography

[1] V Escola do CBPF, Rio de Janeiro, 5 a 16 de julho de 2004. See: http:// mesonpi.cat.cbpf.br/e2004/; *(Super)-Gravities beyond four dimensions*, in Villa de Leyva 2001, Geometric and Topological Methods for Quantum Field Theory. Proceedings of the Summer School Villa de Leyva, Colombia 9–27 July 2001, A. Cardona, S. Paycha, and H. Ocampo, editors. World Scientific, Singapore, (2003). [hep-th:0206169]; *Chern-Simons gravity: From (2+1) to (2n+1) dimensions*, lectures given at the 20th National Meeting of the Physics of Particles and Fields, Sao Lourenco, Brazil, 25–29 Oct 1999 and at the 5th La Hechicera School, Merida, Venezuela, Nov 2000. Braz. J. Phys. **30**, 251–267 (200). [hep-th/0010049].

[2] J. Zanelli, *Lecture Notes on Chern-Simons (Super)-Gravities Second Edition - February 2008*, arXiv:hep-th/0502193v4; *Chern-Simons Forms in Gravitation Theories*, Class. Quant. Grav. **29**, 133001 (2012) [arXiv:1208.3353 [hep-th]]; *Gravitation Theory and Chern-Simons Forms*, Proceedings of the Villa de Leyva School 2011, A. Cardona, S. Paycha, H. Ocampo and Andres Reyes-Lega, editors. World Scientific, Singapore, (2014).

[3] M. Bañados, R. Troncoso and J. Zanelli, *Higher dimensional Chern-Simons supergravity*, Phys. Rev. **D54**, 2605–2611 (1996), [gr-qc/9601003].

[4] R. Troncoso and J. Zanelli, *New gauge supergravity in seven and eleven dimensions*, Phys. Rev. **D58**, R101703 (1998), [hep-th/9710180].

[5] R. Troncoso and J. Zanelli, *Gauge supergravities for all odd dimensions*, Int. Jour. Theor. Phys. **38**, 1181–1206 (1999), [hep-th/9807029].

[6] Matthew D. Schwartz *Quantum Field Theory and the Standard Model*, Cambridge University Press, Cambridge, U. K., (2014).

[7] C. N. Yang and R. Mills, *Conservation of isotpic spin and isotopic gauge invariance*, Phys. Rev. **96**, 191–195 (1954).

[8] M. Nakahara, *Geometry, Topology and Physics*, Boca Raton, USA: Taylor and Francis (2003).

[9] T. Eguchi, P. B. Gilkey, and A. J. Hanson, *Gravitation, gauge theories and differential geometry*, Phys. Rept. **66**, 213–393 (1980).

[10] E. Witten, *(2+1)-Dimensional gravity as an exactly soluble system*, Nucl. Phys. **B311**, 46–78 (1988).

[11] A. Ashtekar and J. Lewandowski, *Background independent quantum gravity: A Status report*, Class. Quant. Grav. **21**, R53 (2004) [gr-qc/0404018].

[12] R. Gambini and J. Pullin, *A First Course in Loop Quantum Gravity*, Oxford University Press (2011).

[13] R. P. Feynman, *Quantum theory of gravitation*, Lecture given at the Conference on Relativistic Theories of Gravitation, Jablonna, Warsaw, Jul 1962. Published in Acta Phys. Polon. **24**, 697–722 (1963).

[14] G. 't Hooft and M. Veltman, *One loop divergencies in the theory of gravitation*, Ann. Inst. H. Poincaré (Phys. Theor.) **A20** (1974) 69–94. G. 't Hooft, *Quantum Gravity: A Fundamental Problem And Some Radical Ideas*, in Proceedings of the 1978 Cargese Summer School, edited by M. Levy and S. Deser (1978), and also in NATO Adv.Study Inst.Ser.B Phys.**44**, 323 (1979).

[15] C.P. Burgess, *Quantum gravity in everyday life: General relativity as an effective field theory*, Living Rev. Rel. **7** (2004) 5, [gr-qc/0311082].

[16] C. Teitelboim, *Quantum mechanics of the gravitational field in asymptotically flat space*, Phys. Rev. **D28**, 310–316 (1983).

[17] M. Bañados, L. J. Garay and M. Henneaux, *Existence of local degrees of freedom for higher dimensional pure Chern-Simons theories*, Phys. Rev. **D53**, R593–596 (1996), [hep-th/9506187]; *The dynamical structure of higher dimensional Chern-Simons theory*, Nucl. Phys. **B476**, 611–635 (1996), [hep-th/9605159].

[18] R. Utiyama, *Invariant theoretical interpretation of interaction*, Phys. Rev. **101**, 1597–1607 (1956).

[19] H. Weyl, *Reine Infinitesimalgeometrie*, Math. Zeit, **2**, 384 (1918); *Eine neue Erweiterung der Relativitätstheorie*, Ann. Phys. (Leipzig) **59**, 101 (1919).

[20] C. Levi-Civita, *Nozione di parallelismo in una varietà qualunque*, Rend. Circ. Mat. Palermo **42** (1917).

[21] S. Weinberg, *The cosmological constant problem*, Rev. Mod. Phys. **61**, 1–23 (1989).

[22] F. London, *Die Theorie von Weyl und die Quanten- mechanik*, Naturwissenschaften **15**, 187; *Quantenmechanische Deutung der Theorie von Weyl*, Z. Phys. **42**, 375.

[23] V. Fock, *Zur Schrödingerschen Wellenmechanik*, Z. Phys. **38**, 242 (1926); *Über die invariante Form der Wellen- und der Bewegungsgleichungen für einen geladenen Massen-punkt*, Z. Phys.**39**, 226 (1926).

[24] J. D. Jackson, and L. B. Okun, *Historical roots of gauge invariance*, Rev. Mod. Phys. **73**, 663 (2001).

[25] P. A. M. Dirac, *The Hamiltonian Form of Field Dynamics*, Canad. J. Math., **3**, 1 (1951); *The Theory of gravitation in Hamiltonian form*, Proc. Roy. Soc. (London) **A246**, 333 (1958);*Fixation of coordinates in the Hamiltonian theory of gravitation*, Phys. rev. **114**, 924 (1959).

[26] P. A. M. Dirac, *Lectures on Quantum Mechanics*, Belfer Graduate School of Science, Yeshiva University, New York, 1964, Dover, New York (2000).

[27] C. Teitelboim, *The Hamiltonian Structure of Spacetime*, Ph. D. Thesis, Princeton Unversity (1973).

[28] Symmetries described by open algebras require a different treatment from the standard one for gauge theories, see, e. g., M. Henneaux, *Hamiltonian form of the path integral for theories with a gauge freedom*, Phys. Rept. **126**, 1–66 (1985).

[29] Euclid, *The Thirteen Books of the Elements* edited by Thomas L. Heath, second Revised edition, Dover, New York (1956).

[30] W. Dunham, *Journey Through Genius: The Great Theorems of Mathematics*, Penguin Books, New York (1991).

[31] R. Debever, *Elie Cartan - Albert Einstein Lettres sur le Parallélisme Absolu, 1929–1932*, Académie Royale de Belgique, Princeton University Press. Princeton (1979).

[32] A. Einstein, *Die Feldgleichungen der Gravitation (The field equations of gravitation)*, Preuss. Akad. Wiss. Berlin, Sitzber. **47**, 844–847 (1915).

[33] D. Hilbert, *Die Grundlagen der Physik*, Konigl. Gesell. d. Wiss. Göttingen, Nachr., Math.-Phys. Kl., **27**, 395–407 (1915).

[34] C. W. Misner, K. S. Thorne and J. A. Wheeler, *Gravitation*, W. H. Freeman, New York (1973).

[35] J. Glanz, *Exploding stars point to a universal repulsive force*, Science **279**, (1998) 651. V. Sahni and A. Starobinsky, *The Case for a Positive Cosmological Lambda Term*, Int. J. Mod. Phys. **D9**, 373–444 (2000), [astro-ph/9904398].

[36] Clifford M. Will, *The confrontation between general relativity and experiment*, Living Rev. Rel. **9**, 3 (2005), [arXiv:gr-qc/0510072].

[37] B. F. Schutz, *A First Course in General Relativity*, Cambridge University Press (1985).

[38] M. Goeckeler and T. Schuecker, *Differential Geometry, Gauge Theories and Gravity*, Cambridge University Press (1987).

[39] R. W. R. Darling, *Differential Forms and Connections*, Cambridge University Press (1994).

[40] Michael Spivak, *Calculus on Manifolds: A Modern Approach to Classical Theorems of Advanced Calculus*, Perseus Pub. Co., Boulder, (1965).

[41] P. D. Alvarez, P. Pais, E. Rodríguez, P. Salgado-Rebolledo and J. Zanelli, *The BTZ black hole as a Lorentz-flat geometry*, Phys. Lett. B **738**, 134 (2014). [arXiv:1405.6657 [gr-qc]].

[42] T. W. B. Kibble, *Lorentz invariance and the gravitational field*, J. Math. Phys. **2**, 212–221 (1961).

[43] C. N. Yang, *Integral formalism for gauge fields*, Phys. Rev. Lett. **33**, 445–447 (1974).

[44] F. Mansouri, *Gravitation as a fauge theory*, Phys. Rev. D **13**, 3192–3200 (1976).

[45] S. W. MacDowell and F. Mansouri, *Unified geometric tTheory of gravity and supergravity*, Phys. Rev. Lett. **38**, 739–742 (1977), Erratum-ibid.**38**, 1376 (1977).

[46] K. Stelle and P.West *Spontaneously broken de Sitter symmetry and the gravitational holonomy group*, Phys. Rev. **D21**, 1466 (1980).

[47] B. Zumino, *Gravity theories in more than four dimensions*, Phys. Rep. **137**, 109–114 (1986).

[48] T. Regge, *On broken symmetries and gravity*, Phys. Rep. **137**, 31–33 (1986).

[49] D. Lovelock, *The Einstein tensor and its generalizations*, J. Math. Phys. **12**, 498–501 (1971).

[50] M. Henneaux, C. Teitelboim and J. Zanelli, *Quantum mechanics for multivalued Hamiltonians*, Phys. Rev. A **36**, 4417 (1987); *Gravity In Higher Dimensions*, proceedings of SILARG V, Rio de Janeiro 1987,edited by M. Novello, World Scientific, Singapore (1987).

[51] See, e.g., http://www-gap.dcs.st-and.ac.uk/history/Mathematicians/Plateau.html

[52] C. Lanczos, *A remarkable property of the Riemann-Christoffel tensor in four dimensions*, Ann. Math. **39**, 842–850 (1938).

[53] B. Zwiebach, *Curvature squared terms and string theories*, Phys. Lett. **B156**, 315–317 (1985).

[54] M. Henneaux, C. Teitelboim and J. Zanelli, *Gauge invariance and degree of freedom count*, Nucl. Phys. **B332**, 169 (1990).

[55] C. Teitelboim and J. Zanelli, *Dimensionally continued topological gravitation theory in hamiltonian form*, Class. and Quantum Grav. **4** (1987) L125–129; and in *Constraint Theory and Relativistic Dynamics*, edited by G. Longhi and L. Lussana, World Scientific, Singapore (1987).

[56] A. Mardones and J. Zanelli, *Lovelock-Cartan theory of gravity*, Class. and Quantum Grav. **8**, 1545–1558 (1991).

[57] H. T. Nieh and M. L. Yan, *An identity in Riemann-Cartan geometry*, J. Math. Phys. **23**, 373 (1982).

[58] O. Chandía and J. Zanelli, *Topological invariants, instantons and chiral anomaly on spaces with torsion*, Phys. Rev. **55**, 7580–7585 (1997), [hep-th/9702025]; *Supersymmetric particle in a spacetime with torsion and the index theorem*, **D58**, 045014 (1998), [hep-th/9803034].

[59] T. Apostol, *Modular Functions and Dirichlet Series in Number Theory*, Springer Verlag, Berlin(1976).

[60] M. Bañados, C. Teitelboim and J. Zanelli, *Lovelock-Born-Infeld theory of gravity*, in J. J. Giambiagi Festschrift, La Plata, May 1990, edited by H. Falomir, R. RE. Gamboa, P. Leal and F. Schaposnik, World Scientific, Singapore (1991).

[61] M. Bañados, C. Teitelboim and J. Zanelli, *Dimensionally continued black holes*, Phys. Rev. **D49**, 975-986 (1994), [gr-qc/9307033].

[62] R. Troncoso and J. Zanelli, *Higher-dimensional gravity, propagating torsion and adS gauge invariance*, Class. and Quantum Grav. **17**, 4451–4466 (2000), [hep-th/9907109].

[63] K. Yano, *Integral Formulas in Riemannian Geometry*, M. Dekker Inc, New York (1970).

[64] J. Crisóstomo, R. Troncoso and J. Zanelli, *Black Hole Scan*, Phys. Rev. **D62**, 084013 (2000), [hep-th/0003271].

[65] R. Aros, M. Contreras, R. Olea, R. Troncoso and J. Zanelli, *Conserved charges for gravity with locally AdS asymptotics*, Phys. Rev. Lett. **84**, 1647 (2000), [gr-qc/9909015]; *Conserved charges for even dimensional asymptotically AdS gravity theories*, Phys. Rev. D **62**, 044002 (2000), [hep-th/9912045].

[66] J. Zanelli, *Quantization of the gravitational constant in odd dimensions*, Phys. Rev. **D51**, 490–492 (1995), [hep-th/9406202].

[67] P. A. M. Dirac, *Quantised Singularities in the Electromagnetic Field*, Proc. Roy. Soc. (London) **A133**, 60 (1931).

[68] R. Gilmore, *Lie Groups Lie Algebras and Some of Their Applications*, Wiley, New York (1974).

[69] E. Inönü, *A historical note on group contractions*, in http://ysfine.com/wigner/inonu.pdf, Feza Gursey Institute, Istambul, (1997).

[70] A. Achúcarro and P. K. Townsend, *A Chern-Simons action for three-dimensional anti-de Sitter supergravity theories*, Phys. Lett. **B180**, 89 (1986).

[71] E. Witten, *(2+1)-Dimensional Gravity as an Exactly Soluble System*, Nucl. Phys. B **311**, 46 (1988).

[72] E. Witten, *Three-Dimensional Gravity Revisited*, arXiv:0706.3359 [hep-th].

[73] S.S. Chern and J. Simons, *Characteristic Forms and Geometric Invariants*, Ann. Math. **99**, 48 (1974).

[74] J. Zanelli, *Uses of Chern-Simons actions*, AIP Conf. Proc. **1031**, 115 (2008); *Ten Years of AdS/CFT Conjecture*, Buenos Aires, December 2007. [arXiv:0805.1778 [hep-th]].

[75] A. Chamseddine, *Topological gauge theory of gravity in five dimensions and all odd dimensions*, Phys. Lett. **B233**, 291–294 (1989); *Topological gravity and supergravity in various dimensions*, Nucl. Phys. **B346**, 213–234 (1990).

[76] F. Müller-Hoissen, *From Chern-Simons to Gauss-Bonnet*, Nucl. Phys. **B346**, 235–252 (1990).

[77] P. Mora, R. Olea, R. Troncoso and J. Zanelli, *Finite action principle for Chern-Simons adS gravity*, JHEP **0406**, 036 (2004), [hep-th/0405267].

[78] S. A. Hartnoll, *Lectures on holographic methods for condensed matter physics*, Class. Quant. Grav. **26** (2009) 22400, [arXiv:0903.3246 [hep-th]].

[79] G. T. Horowitz, *Introduction to Holographic Superconductors*, arXiv:1002.1722 [hep-th].

[80] S. A. Hartnoll, C. P. Herzog and G. T. Horowitz, *Holographic Superconductors*, JHEP **0812**, 015 (2008).

[81] C. P. Herzog, *Lectures on Holographic Superfluidity and Superconductivity*, J. Phys. A **42** (2009) 343001. [arXiv:0904.1975 [hep-th]].

[82] A. H. MacDonald, in les Houches, Session LXI, 1994, Physique Quantique Meso-scopique, edited by E. A. Kermans, G. Montambeaux and J. L. Pichard /Elsevier, Amsterdam, 1995.

[83] S. Rao, *An anyon primer*, hep-th/920966.

[84] The quantum Hall Effect, ed. R. E. Prange and S. M. Girvin, Springer Verlag, New York (1986).

[85] T. L. Hughes, X.-L. Qi and S. C. Zhang, Phys. Rev. B 78, 195424 (2008), and references therein. See also X.-L. Qi and S. C. Zhang, Rev. Mod. Phys. 83, 1057 (2011) and references therein.

[86] A. H. Castro Neto, F. Guinea, N. M. R. Peres, K. S. Novoselov, A. K. Geim, Rev. Mod. Phys. **81** (2009) 109.

[87] F. Guinea, M. I. Katznelson and M. A. H. Vozmediano, *Gauge fields in graphene*, Phys. Rept. (2010), doi:10.1016/j.physrep.2010.07.003.

[88] A. Iorio and G. Lambiase, *Quantum field theory in curved graphene space-times, Lobachevsky geometry, Weyl symmetry, Hawking effect and all that*, ...[arXiv:1308.0265]

[89] D. G. Boulware and S. Deser, *String Generated Gravity Models*, Phys. Rev. Lett. **55**, 2656 (1985).

[90] C. Teitelboim and J. Zanelli, *Dimensionally continued topological gravitation theory in Hamiltonian form*, Class. Quant. Grav. **4**, L125 (1987).

[91] P. Mora, R. Olea, R. Troncoso and J. Zanelli, *Transgression forms and extensions of Chern-Simons gauge theories*, JHEP **0601**, 067 (2006), [hep-th/0601081].

[92] J.Zanelli, *Chern-Simons forms and transgression actions, or the universe as a sub-system*, in Proceedings of the 12th NEB Meeting, Recent Developments in Gravity, Nafplio, Greece, June 2006. J. Phys. Conf. Ser. **68**: 012002 (2007).

[93] O. Mišković and R. Olea, *On boundary conditions in three-dimensional adS gravity*, Phys. Lett. **B640**, 101-107 (2006), [hep-th/0603092]; *Counterterms in dimensionally continued adS gravity*, JHEP **0710**, 028 (2007), [arXiv: 0706.4460].

[94] C. Fefferman and R. Graham, *Conformal invariants*, in The mathematical heritage of Elie Cartan (Lyon 1984), Astérisque, 1985, Numero Hors Serie, 95.

[95] M. Bañados, O. Mišković and S. Theisen, *Holographic currents in first order gravity and finite Fefferman-Graham expansions*, JHEP **0606** (2006) 025, [hep-th/0604148].

[96] S. Coleman and J. Mandula *All possible symmetries of the S matrix*, Phys. Rev. **159**, 1251–1256 (1967).

[97] S. Deser, R. Jackiw and G. 't Hooft, *Three-Dimensional Einstein Gravity: Dynamics of Flat Space*, Annals Phys. **152**, 220 (1984).

[98] S. Deser and R. Jackiw, *Three-Dimensional Cosmological Gravity: Dynamics of Constant Curvature*, Annals Phys. **153**, 405 (1984).

[99] B. Reznik, *Thermodynamics of event horizons in (2+1)-dimensional gravity*, Phys. Rev. D **45**, 2151 (1992).

[100] S. Carlip, *Quantum Gravity in 2+1 Dimensions*, Cambridge U. Press (1998).

[101] M. Bañados, C. Teitelboim and J. Zanelli, *Black hole in three-dimensional spacetime*, Phys. Rev. Lett. **69**, 1849 (1992), [hep-th/9204099].

[102] M. Bañados, M. Henneaux, C. Teitelboim and J. Zanelli, *Geometry of the (2+1) Black Hole*, Phys. Rev. **D48**, 1506 (1993), [gr-qc/9302012].

[103] C. Martínez, C. Teitelboim and J. Zanelli, *Charged rotating black hole in three spacetime dimensions*, Phys. Rev. D **61**, 104013 (2000) [hep-th/9912259].

[104] N. Cruz, C. Martínez and L. Peña, *Geodesic structure of the (2+1) black hole*, Class. Quant. Grav. **11**, 2731 (1994) [gr-qc/9401025].

[105] S. F. Ross and R. B. Mann, *Gravitationally collapsing dust in (2+1)-dimensions*, Phys. Rev. D **47**, 3319 (1993) [hep-th/9208036].

[106] Y. Peleg and A. R. Steif, Phase transition for gravitationally collapsing dust shells in (2+1)-dimensions, Phys. Rev. D **51**, 3992 (1995) [gr-qc/9412023].

[107] D. Birmingham and S. Sen, Gott time machines, BTZ black hole formation, and Choptuik scaling, Phys. Rev. Lett. **84**, 1074 (2000) [hep-th/9908150].

[108] N. Cruz and J. Zanelli, *Stellar equilibrium in (2+1)-dimensions*, Class. Quant. Grav. **12**, 975 (1995) [gr-qc/9411032].

[109] James T. Wheeler, *Symmetric Solutions to the Gauss-Bonnet Extended Einstein Equations*, Nucl. Phys. B **268**, 737 (1986). *Symmetric Solutions to the Maximally Gauss-Bonnet Extended Einstein Equations*, Nucl. Phys. B **273**, 732 (1986).

[110] Brian Whitt, *Spherically Symmetric Solutions of General Second Order Gravity*, Phys. Rev. D **38**, 3000 (1988).

[111] Robert C. Myers and Jonathan Z. Simon, *Black Hole Thermodynamics in Lovelock Gravity*, Phys. Rev. D **38**, 2434 (1988).

[112] David L. Wiltshire, *Black Holes in String Generated Gravity Models*, Phys. Rev. D **38**, 2445 (1988).

[113] J. Earman, *Bangs, Crunches, Whimpers, and Shrieks: Singularities and Acausalities in Relativistic Space-Times* Oxford University Press, New York (1995).

[114] O. Mišković and J. Zanelli, *On the negative spectrum of the 2+1 black hole*, Phys. Rev. D **79**, 105011 (2009) [arXiv:0904.0475 [hep-th]].

[115] R. Jackiw, in Proceedings at the SILARG VII, Cocoyoc, Mexico, 1990; also in *Diverse Topics in Theoretical and Mathematical Physics*, World Scientific, Singapore (1995).

[116] H. J. Matschull, *Black hole creation in (2+1)-dimensions*, Class. Quant. Grav. **16**, 1069 (1999) gr-qc/9809087].

[117] M. M. Müller, M. Ben Amar, and J. Guven, *Conical Defects in Growing Sheets*, Phys. Rev. Lett. **101**, 156104 (2008). http://physics.aps.org/story/v22/st12

[118] O. Miskovic and J. Zanelli, *Couplings between Chern-Simons gravities and 2p-branes*, Phys. Rev. D **80**, 044003 (2009) [arXiv:0901.0737 [hep-th]].

[119] J. D. Edelstein, A. Garbarz, O. Miskovic and J. Zanelli, *Stable p-branes in Chern-Simons AdS supergravities*, Phys. Rev. D **82**, 044053 (2010) [arXiv:1006.3753 [hep-th]].

[120] O. Mišković and J. Zanelli, *Couplings between Chern-Simons gravities and 2p-branes*, Phys. Rev. D **80**, 044003 (2009) [arXiv:0901.0737 [hep-th]].

[121] J. D. Edelstein, A. Garbarz, O. Miskovic and J. Zanelli, *Naked Singularities, Topological Defects and Brane Couplings*, Int. J. Mod. Phys. D **20**, 839 (2011) [arXiv:1009.4418 [hep-th]].

[122] J. D. Edelstein, A. Garbarz, O. Miskovic and J. Zanelli, *Geometry and stability of spinning branes in AdS gravity*, Phys. Rev. D **84**, 104046 (2011) [arXiv:1108.3523 [hep-th]].

[123] S. Coleman, *Uses of instantons*, Erice Lectures 1977. A. Zichcchi, ed. Reprinted in *Aspects of Symmetry*, Cambridge University Press (1985).

[124] L. Huerta and J. Zanelli, *Optical Properties of a θ-Vacuum* Phys. Rev. D **85**, 085024 (2012)[arXiv:1202.2374 [hep-th]].

[125] A. Anabalón, S. Willison and J. Zanelli, *The Universe as a topological defect*, Phys. Rev. D **77**, 044019 (2008) [hep-th/0702192].

[126] S. P. Martin, *A Supersymmetry primer*, Adv. Ser. Direct. High Energy Phys. **21**, 1 (2010) [hep-ph/9709356].

[127] P. van Nieuwenhuizen, *Supergravity*, Phys. Rep. **68**, 189–398 (1981).

[128] S. Weinberg, *The Quantum Theory of Fields*, Vol III, Cambridge University Press, Cambridge U.K. (200).

[129] P. G. O. Freund, *Introduction to Supersymmetry*, Cambridge University Press (1988).

[130] M. Sohnius, *Introducing Supersymmetry*, Phys. Rept. **28**, 39–204 (1985).

[131] J.W. van Holten and A. Van Proeyen, *N=1 supersymmetry algebras in D=2, D=3, D=4 mod-8*, J. Phys. **A15**, 3763–3779 (1982).

[132] V. G. Kac, *A sketch of Lie superalgebra theory*, Comm. Math. Phys. **53**, 31–64 (1977).

[133] P. K. Townsend, *Three lectures on quantum sSupersymmetry and supergravity*, Supersymmetry and Supergravity '84, Trieste Spring School, April 1984, B. de Wit, P. Fayet, and P. van Nieuwenhuizen, editors, World Scientific, Singapore (1984).

[134] V. O. Rivelles and C. Taylor, *Off-Shell extended supergravity and central charges*, Phys. Lett. **B104**, 131–135 (1981); *Off-Shell No-Go Theorems for Higher Dimensional Supersymmetries and Supergravities*, Phys. Lett. **B121**, 37–42 (1983).

[135] A. Pankiewicz and S. Theisen, *Introductory lectures on string theory and the adS/CFT correspondence*, in: Villa de Leyva 2001, Geometric and topological methods for quantum field theory. Proceedings of the Summer School Villa de Leyva, Colombia 9-27 July 2001. A. Cardona, S. Paycha, and H. Ocampo, editors. World Scientific, Singapore, (2003).

[136] S. Deser, B. Zumino *Consistent supergravity*, Phys. Lett. **B62**, 335 (1976). S. Ferrara, D. Z.Freedman, and P. van Nieuwenhuizen, *Progress toward a theory of supergravity*, Phys. Rev. **D13**, 3214 (1976). D. Z. Freedman, and P. van Nieuwenhuizen, *Properties of supergravity theory*, Phys. Rev. **D14**, 912 (1976).

[137] A. Losev, M. A. Shifman, A. I. Vainshtein, *Counting supershort supermultiplets*, Phys. Lett. **B522**, 327–334 (2001), [hep-th/0108153].

[138] C. Fronsdal, *Elementary particles in a curved space. II*, Phys. Rev. **D10**, 589–598 (1974).

[139] A. Salam and E. Sezgin, *Supergravity in Diverse Dimensions*, World Scientific, Singapore (1989).

[140] P.K. Townsend, *Cosmological constant in supergravity*, Phys. Rev. **D15**, 2802-2804 (1977).

[141] E. Cremmer, B. Julia, and J. Scherk, *Supergravity theory in eleven dimensions*, Phys. Lett. **B76**, 409–412 (1978).

[142] K.Bautier, S.Deser, M.Henneaux and D.Seminara, *No cosmological D=11 supergravity*, Phys. Lett. **B406**, 49–53 (1997) [hep-th/9704131].

[143] S. Deser, *Uniqueness of d = 11 supergravity*, in Black Holes and the Structure of the Universe, C. Teitelboim and J. Zanelli, editors, World Scientific, Singapore (2000), [hep-th/9712064].

[144] J. Milnor, *Morse Theory.* Princeton University Press (1969).

[145] O. Chandía, R. Troncoso and J. Zanelli, *Dynamical content of Chern-Simons supergravity*, Second La Plata Meeting on Trends in Theoretical Physics, Buenos Aires, (1998), H. Falomir, R.E.Gamboa Saraví and F.A.Schaposnik, editors, American Institute of Physics (1999) [hep-th/9903204].

[146] R. Troncoso, *Supergravedad en Dimensiones Impares*, Doctoral Thesis, Universidad de Chile, Santiago (1996).

[147] P. Horava and E. Witten, *Heterotic and type I string dynamics from eleven-dimensions*, Nucl. Phys. B **460**, 506 (1996) [hep-th/9510209].

[148] E.P. Wigner and E. Ínönü, Proc. Nat-Acad . Sci 39, 510-524 (1953).

[149] I. E. Segal, *A class of operator algebras which are determined by groups*, Duke Math. J. **18**, 221 (1951).

[150] J. A. de Azcárraga, J. M. Izquierdo, M. Picón and O. Varela, *Generating Lie and Gauge Free Differential (Super)Algebras by Expanding Maurer-Cartan Forms and Chern-Simons Supergravity* Nucl. Phys. **B662**, (2003) 185, [hep-th/021234].

[151] M. Hatsuda and M. Sakaguchi, *Wess-Zumino Term for the AdS Superstring and Generalized Inönü-Wigner Contraction* Prog. Theor. Phys. **109**, (2003) 853, [hep-th/0106114].

[152] M. Hassaine and M. Romo, *Local supersymmetric extensions of the Poincare and AdS invariant gravity*, JHEP **0806**, 018 (2008), [arXiv:0804.4805 [hep-th]].

[153] M. Hassaïne, R. Troncoso and J. Zanelli, *Poincaré Invariant Gravity with Local Supersymmetry as a Gauge Theory for the M-Algebra*, Phys. Lett. **B596**, (2004) 132, [hep-th/0306258].

[154] K. Popper, *The Logic of Scientific Discovery*, Hutchinson (London, 1959).

[155] P. D. Alvarez, M. Valenzuela and J. Zanelli, *Supersymmetry of a different kind*, JHEP **1204**, 058 (2012) [arXiv:1109.3944 [hep-th]].

[156] P. D. Alvarez, P. Pais and J. Zanelli, *Unconventional supersymmetry and its breaking*, Phys. Lett. B **735**, 314 (2014) [arXiv:1306.1247 [hep-th]].

[157] J. Zanelli, *2+1 black hole with SU(2) hair (and the theory where it grows)*, in Spanish Relativity Meeting: Almost 100 years after the Einstein Revolution (ERE 2014), J. Phys. Conf. Ser. **600**, no. 1, 012005 (2015).

[158] P. D. Alvarez, P. Pais, E. Rodríguez, P. Salgado-Rebolledo and J. Zanelli, *Supersymmetric 3D model for gravity with SU(2) gauge symmetry, mass generation and effective cosmological constant* (2015, unpublished) [arXiv:1505.03834 [hep-th]].

Index

adjoint representation, 6, 117, 118
affine connection, 15
affine structure, 14, 16
affinity, 11
Albert Einstein, 15
algebra of diffeomorphisms, 9
angular deficit, 13
angular excess, 74
anomalies, 2
anti-de Sitter group, 34
anticommutator, 83
apex of a cone, 13
auxiliary fields, 85

Bianchi identity, 30
black hole, 69
Bose-Einstein statistics, 82
boson, 82
Bott periodicity, 95

carriers of interactions, 1
Cartan's point of view, 15
Cauchy data, 20
central charge, 97
change of coordinates, 14
Characteristic 2n-form, 56
characteristic class, 56
Chern character, 115
Chern class, 42
Chern-Simons forms, 56
chirality, 89
Christoffel symbol, 7, 15, 24, 26
chromodynamics, 6
compass, 11
conical singularity, 73
connection, 5, 6, 14–16

connection Γ, 17
conserved charge, 49
contraction, 107
contorsion tensor, 29
coordinate change, 7
coordinate transformation, 4
covariant derivative, 6, 7, 14, 16, 17
covariant derivative operator, 7
curvature, 13
curvature of a manifold, 13
curvature two-form, 16

de Sitter group, 34
diffeomorphism group, 15, 17
diffeomorphisms, 4, 5, 24
Dirac algebra, 88
Dirac matrices, 87, 96, 97, 119
Dirac spinors, 101
Dirac's quantization, 48
Dirichlet boundary, 63
Dirichlet problem, 66

effective theory, 4
Einstein equations, 19
Einstein-Hilbert action, 21, 33
electroweak, 9
Elie Cartan, 15
Euclid, 11
Euclidean geometry, 11
Euler characteristic, 37, 38
Euler density, 52
exotic AdS Lagrangian, 95
exotic Lagrangian, 53, 59

Fermi-Dirac statistics, 82
fermion, 15, 82

Printed in the United States
By Bookmasters